Generalized Filter Design
by Computer Optimization

For a complete listing of the *Artech House Microwave Library*,
turn to the back of this book.

Generalized Filter Design by Computer Optimization

Djuradj Budimir

Artech House
Boston • London

Library of Congress Cataloging-in-Publication Data
Budimir, Djuradj
    Generalized filter design by computer optimization / Djuradj Budimir
    p.   cm. — (Artech House microwave library)
    Includes bibliographical references and index.
    ISBN 0-89006-579-9 (alk. paper)
    1. Electric filters—Design and construction—Data processing.
    2. Computer-aided design. I. Title. II. Series.
TK7872.F5B82 1998
621.3815'324'028551—dc21   97-39465
                                                                        CIP

British Library Cataloguing in Publication Data
Budimir, Djuradj
    Generalized filter design by computer optimization
    (Artech House microwave library)
    1. Electric filters–Design and construction–Data processing
    I. Title
    621.3'815324'0285

    ISBN 0-89006-579-9

Cover design by Jennifer L. Stuart

© 1998 ARTECH HOUSE, INC.
685 Canton Street
Norwood, MA 02062

International Standard Book Number: 0-89006-579-9
Library of Congress Catalog Card Number: 97-39465

10 9 8 7 6 5 4 3 2 1

To
Aleksandar and Slobodan

Contents

Preface

Recent advances in microwave computer-aided filter design technology suggest the feasibility of interfacing electromagnetic simulations directly to sophisticated optimization systems. With the availability of powerful computers and RISC workstations, this optimization-based approach to the design of microwave filters becomes a desirable tool.

The central theme of this book is the optimization-oriented approach for the accurate design of radio frequency, microwave, and millimeter-wave filters. It presents computer-aided filter design algorithms and provides examples of their applications. This book is divided into nine chapters. Chapter 1 gives an introduction. Chapter 2 considers transmission lines, lumped elements, and resonators. Chapter 3 describes the characterization of discontinuities. Chapter 4 presents an optimization method used in filter design. Chapters 5 through 8 concentrate on the design of lumped element, E-plane metal insert, ridged waveguide, and coplanar waveguide filters. The final chapter describes computer-aided filter design programs.

This book is intended for advanced undergraduate and graduate courses, for design engineers, and for research and development specialists who are interested in computer-aided filter design of radio frequency, microwave, and millimeter-wave filters. The material represents state-of-the-art technology at the time of writing. The reader is assumed to have completed courses at the basic level in filter theory, electromagnetics, and microwaves.

The material in this book is based on my own work at the University of Leeds and the University of London over the last seven years. I wish to give particular thanks to Professor J. David Rhodes of Filtronic Components and the Department of Electronic and Electrical Engineering, University of Leeds, who provided expert advice throughout the course of this work. Further thanks are due to Dr. Vasil Postoyalko, Dr. John R. Richardson and Dr. Stavros Iezekiel for their encouragement and during part of this work, Professor Roger Pollard for useful technical discussions and assisting in the microwave measurement, and Mr. Terry Moseley for expert machining of the waveguide housing, all of whom are at the Department of Electronic and Electrical Engineering, University of Leeds. Furthermore, I would like to acknowledge Professor Alec Cullen, Department of Electronic and Electrical Engineering, University College, University of London, for useful technical suggestions. Thanks are also due to Dr. Ian Robertson for permission to undertake part of this work in the Department of Electronic and Electrical Engineering, King's College, University of London.

1

Introduction to Computer-Aided Filter Design

Computer-aided design (CAD) of *radio frequency* (RF), microwave, and *millimeter-wave* (mm-wave) filter structures must be capable of handling the restrictions of a widespread application of low-cost precision fabrication methods, such as computer-controlled milling, spark eroding, or photolithographic etching techniques, in which postassembly tuning is no longer economical or feasible. The filter design also must meet the demands of the expanding utilization of higher frequency bands (up to mm-waves), which need tighter tolerances. These conditions require CAD, rigorous electromagnetic simulation techniques, and efficient computer optimization methods that allow the computer-aided filter design to take into account all the significant design parameters.

CAD strictly interpreted can be taken to mean any design process where the computer is used as a tool. However, usually the term CAD implies that without the computer as a tool, that particular design process would have been impossible or much more difficult, more expensive, and more time consuming. A typical flow diagram for circuit CAD is shown in Figure 1.1. To implement a conventional design procedure to meet the given filter specifications successfully, a CAD approach becomes necessary. When the CAD approach is used, experimental modifications of the filter (which is unavoidable in the conventional design procedure) are replaced by a computer-based optimization of the initial design.

A typical design process usually begins with a given set of specifications or design goals for the filter. Synthesis methods and available design data help to arrive at the initial filter design. The performance of the initial filter design is evaluated by computer-aided filter analysis. Numerical methods needed for

1

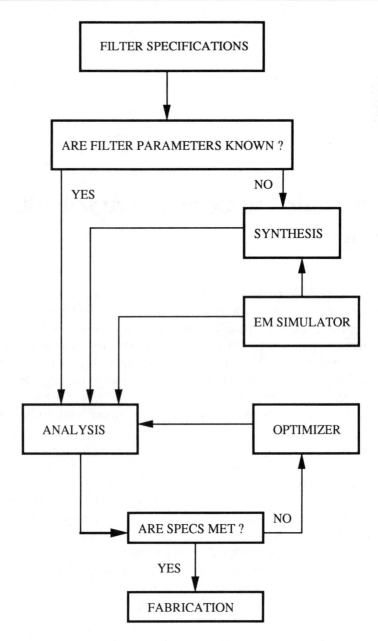

Figure 1.1 Flow chart for a CAD procedure.

the analysis of filter structures are called from the library of subroutines developed for that purpose. Filter characteristics obtained as a result of the analysis are compared with the given specifications. If the results fail to satisfy the desired specifications, the designable parameters of the filter are altered in a systematic manner. The optimization is especially important for mm-wave filters because they cannot easily be tuned or trimmed after fabrication. The sequence of filter analysis, comparison with the desired performance, and parameter modification is performed iteratively until the specifications are met or the optimum performance of the filter is achieved. If the specifications are not met, the design may have to be revised. When the design satisfies the specifications, a filter can be built and tested. If the measured results meet the specifications, the design process is completed; otherwise, the design needs to be repeated. The filter then is fabricated and experimental measurements are carried out. Some modifications may still be required if the modeling has not been accurate enough. Ideally, the modifications will be very small, and the aim of the CAD method is to minimize the experimental iterations as far as practicable. Thus, CAD can greatly decrease the time and the cost of design while enhancing its quality.

The process of CAD, as outlined in the preceding paragraphs, consists of three important segments:

- Electromagnetic simulation;
- Computer-aided analysis;
- Computer optimization.

1.1 Electromagnetic Simulation

Electromagnetic simulation involves the characterization of various filter discontinuities, to the extent of providing a numerical model that can be handled by the computer. An accurate and reliable characterization of filter discontinuities is one of the basic prerequisites of successful CAD. The degree of accuracy to which the performance of filters can be predicted depends on the accuracy of the characterization and simulating of those discontinuities. S parameters are convenient for characterization of filter structures. Detailed numerical simulating of filter structures becomes involved and time consuming. Characterization and simulating of various types of waveguide (rectangular and ridged) discontinuities such as step discontinuity rectangular waveguide to metal septum in rectangular waveguide and step discontinuity metal septum in rectangular waveguide to ridged waveguide by the mode-matching method are discussed in Chapter 3. The effects of those discontinuities become more and more significant as one moves from the microwave frequency range to mm-waves with the need for lower tolerances. Such effects require rigorous electromagnetic simulation methods that allow the accurate computer-aided component design to take into account all significant factors, such as finite thickness of septa. Much research aimed at characterization for CAD purposes has been reported for waveguide discontinuities [1–5]. Sonnet software's Em™ package uses the method of moments used for electromagnetic simulation of various types of coplanar waveguide discontinuities, as discussed in Chapter 3 [6]. Table 1.1 summarizes some commercially available electromagnetic simulators for computer-aided filter design purposes [6–15].

1.2 Computer-Aided Analysis

Computer-aided analysis constitutes the key step in the CAD procedure. Because the analysis forms a part of the optimization loop, the analysis subpro-

Table 1.1
Some Commercially Available Electromagnetic Simulators

Company	Product*	Type
HP-EEsof (HP range)	Momentum	3D planar electromagnetic
	HFSS	3D arbitrary electromagnetic
Compact Software	Microwave Explorer	3D planar electromagnetic
Sonnet Software	Em	3D planar electromagnetic
	Xgeom	Layout entry
	Emvu	Current display
Jansen Microwave	Unisym/Sfpmic	3D planar electromagnetic
Ansoft Corporation	Maxwell Strata	3D planar electromagnetic
	Maxwell Eminence	3D arbitrary
	MicroWaveLab	3D arbitrary electromagnetic
ArguMens	Stingray	3D planar electromagnetic
Zeland Software	IE3D	3D planar electromagnetic
Optimization System Associates	Empipe	Sonnet, Ansoft, HP optimization
	Empipe3D/EmpipeExpress	Geometry capture front end
MacNeal-Schwendler Corp.	MicroWaveLab	3D arbitrary electromagnetic
Kimberley Communications Consultants	Micro-Stripes	3D arbitrary electromagnetic (TLM)
Boulder Microwave Tech.	Ensemble	3D planar
Computer System Technologies	Mafia	3D planar

*All trademarks acknowledged.

gram is executed again and again for a specific filter design. For that reason, an efficient analysis algorithm constitutes the backbone of any CAD package. Computer-aided analysis provides the response of a specified filter configuration to a given set of inputs. Computer-aided analysis is perhaps the most developed and most widely used aspect of CAD. The filters considered in this book can be expressed as a cascade combination of two-port sections. The analysis problem in that case may be stated as follows: Characterize the filter sections (say, in terms of individual scattering, ABCD matrices, or even or odd mode impedances) and find the scattering, ABCD matrices, or, by choice, even and odd mode impedances of the overall filter structure. Computer-aided analysis is the process of evaluating the filter performance, as follows:

1. The lumped-element lowpass filter is decomposed into building blocks, such as inductor and capacitor, while the conventional E-plane bandpass filter structure is decomposed into appropriate key building blocks, such as homogeneous rectangular waveguide and step discontinuity rectangular waveguide to metal septum in rectangular waveguide.

2. The ridged waveguide bandpass filter structure is decomposed into building blocks such as homogeneous rectangular waveguide, step discontinuity rectangular waveguide to metal septum in rectangular waveguide, step discontinuity metal septum in rectangular waveguide to ridged waveguide, and the ridged waveguide.

3. Meanwhile, the coplanar waveguide bandpass filter structure is decomposed into building blocks such as coplanar waveguide, edge coupled coplanar waveguide discontinuity, and end-coupled coplanar waveguide discontinuity.

4. Microwave analysis of the conventional E-plane bandpass filter structures involves evaluation of even and odd mode impedances of the overall filter structure in terms of the given even and odd mode impedances of individual sections of the filter, while microwave analysis of the lumped-element, ridged-waveguide and coplanar-waveguide filter structures involves evaluation of ABCD parameters of the overall filter structure in terms of the given ABCD parameters of individual sections of the filter.

1.3 Computer Optimization

An important class of problems in the CAD of microwave filters concerns algorithms for adjusting circuit parameters to minimize the deviations between the circuit performance achieved at some stage of the design and the desired specifications. As shown in the flow diagram in Figure 1.1, one starts with a given set of filter specifications and an initial filter design. Filter characteristics obtained from the analysis are compared with the given specifications. If the results fail to satisfy the desired specifications, the designable parameters of the filter are altered in a systematic manner. The sequence of filter analysis, comparison with the designed performance, and parameter modification is performed iteratively until the optimum performance of the filter is achieved. The process is termed optimization. Most optimization techniques employed in RF, microwave, and mm-wave CAD employ general forms of error minimization algorithms [16–25] (Table 1.2). The accuracy of the algorithms is related to the number of discrete frequency points used to define the error vector. Cohn introduced an optimization algorithm that requires fewer sampling points to achieve convergence than generalized error minimization algorithms [26].

The intention here is to provide the reader with a numerical optimization method [27], based on Cohn's, that gives an accurate equal-ripple response within the passband for filters based on Chebyshev, generalized Chebyshev, and elliptic function prototypes. It is desirable to have good starting values (filter dimensions) for this optimization method. Using the conventional synthesis procedures, starting values of the optimization routine can be obtained. The advantages of this method are that the problems of local minima are avoided, it requires fewer sampling points to achieve convergence than generalized error

Table 1.2
Some Commercially Available Filter Optimization Tools

Company	Product*	Type
HP-EEsof (HP range)	MDS	Integrated package
	Series IV/PC	Integrated package
HP-EEsof (EEsof range)	Touchstone	Linear
	Linecalc	Physical parameters
	E-Syn	Filter synthesis
Compact Software	SuperCompact	Linear
Jansen Microwave	Linmic+	Linear (integrated suite)
ArguMens	Octopus	Linear
Optimization System Associates	Hope	Linear/nonlinear
	Empipe	Sonnet, Ansoft, HP optimization
	Empipe3D/EmpipeExpress	Geometry capture front end
Optotek	MMICAD	Linear
Eagleware	M/FILTER	Linear

*All trademarks are acknowledged.

minimization algorithms, and the Chebychev criteria are satisfied [28]. This method can handle symmetrical and asymmetrical lowpass, highpass, and bandpass Chebyshev, generalized Chebyshev, and elliptic filters [26,27,29,30–34]. To illustrate the application of the method described in this book, the designs of lumped-element lowpass, E-plane metal-insert bandpass, ridged-waveguide bandpass, and coplanar waveguide bandpass filters are considered as examples. When implemented around electromagnetic simulators, the method can be used to include all the effects of discontinuities, junctions, and so on, to reduce the amount of tuning required in the final filter.

1.4 Outline of This Book

The need to develop an efficient and accurate method for the design of RF, microwave, and mm-wave filters has already been discussed. It is the main aim of this book to make a contribution to this area. Key stages in the filter design are addressed, namely, electromagnetic simulation of filter discontinuities, determination of the starting points for the optimization algorithm, and the development of a numerical optimization algorithm. Toward that end, Chapter 2 discusses characterization of various types of transmission lines, lumped elements, and resonator structures. Chapter 3 deals with the electromagnetic analysis of the discontinuities in E-plane metal-insert filter structures such as the metal septum in rectangular waveguide and the metal septum in rectangular waveguide between two ridged waveguides with equal and different gaps by the rigorous full-wave mode-matching method. In the case of longitudinally symmetrical filter structures such as conventional E-plane metal insert filters, we need consider only the structures with the electric and magnetic walls at the plane of symmetry (half-filter structure). Each E-plane septum is itself symmetrical and can be electrically represented by normalized even and odd mode impedances. The mode-matching method, where up to 140 modes may be included, has been used for calculation of the even and odd mode impedances. Analysis of ridged waveguide by the generalized transverse resonance method is described. The convergence mechanism has been studied to achieve a reliable solution with minimal numerical computation.

Chapter 4 presents an equal-ripple approach to numerical optimization of E-plane metal-insert filters. The advantages of the proposed optimization method over generalized methods available in software packages like EEsof Touchstone™ [13] and Compact Software SuperCompact™ [35] are discussed. Chapter 5 presents the design of lumped element filters by computer optimization. Chapter 6 deals with a new approach to the design of conventional E-plane metal-insert bandpass filters by equal-ripple optimization. Determination of the starting point for the optimization algorithm is described. The analysis of those filters has neglected higher-order mode coupling between E-plane septa. For the design example considered, that was shown not to be important. When higher-order mode coupling needs to be taken into account, the numerically efficient procedure for the cascading of filter sections, which involves only real scalar arithmetic, no longer can be applied. The optimization aspect of the presented design method is still applicable. To confirm the accuracy of the design procedure, a five-resonator X-band conventional E-plane bandpass filter was fabricated. Specific characteristics of the filter are discussed and measurement results of given filter responses presented.

Chapter 7 examines the design of ridged-waveguide bandpass filters by computer optimization. The modified design procedure, which should include the concept of impedance inverters and impedance scaling of the impedance levels of the prototype filter for design of ridged-waveguide bandpass filters, is presented. A five-resonator X-band E-plane ridged waveguide bandpass filter is designed and fabricated. Chapter 8 describes the design of coplanar waveguide bandpass filters by computer optimization. Finally, Chapter 9 describes CAD programs.

References

[1]   Itoh, T., *Numerical Techniques for Microwave and Millimeter-Wave Passive Structures,* New York: Wiley, 1989.

[2]   Collin, R. E., *Field Theory of Guided Waves,* New York: McGraw-Hill, 1960.

[3]   Schwinger, J., and D. Saxon, *Discontinuities in Waveguide (Documents on Modern Physics),* New York: Gordon and Greach, 1968.

[4]   Wexler, A., "Solution of Waveguide Discontinuities by Modal Analysis," *IEEE Trans. Microwave Theory & Tech.,* Vol. MTT-15, September 1967, pp. 508–517.

[5]   Mittra, R., and S. W. Lee, *Analytical Techniques in the Theory of Guided Waves,* New York: Macmillan, 1971.

[6]   *Em User's Manual,* Vol. 1, Release 4.0, Sonnet Software Inc., Liverpool, NY, 1996.

[7]   *HFSS Reference Manual,* Release 2.0, Hewlett-Packard Co., Palo Alto, CA, 1992.

[8]   *Micro-Stripes User's Manual,* Release 2.3, Kimberley Communications Consultants Ltd., Nottingham, England, 1994.

[9]   *Maxwell Strata User's Manual,* Ansoft Corp., Pittsburgh, PA, 1996.

[10] *Maxwell Eminence User's Manual,* Ansoft Corp., Pittsburgh, PA, 1996.

[11] *MicroWaveLab User's Manual,* Ansoft Corp., Pittsburgh, PA, 1996.

[12] *MicroWaveLab User's Manual,* MacNeal-Schwendler Corp., Milwaukee, WI, 1996.

[13] *Mafia User's Manual,* Version 3.2, Computer Simulation Technology, Darmstadt, Germany, 1995.

[14] *IE3D User's Manual,* Zeland Software Inc., Fremont, CA, 1995.

[15] *Microwave Explorer User's Manual,* Compact Softwaree Inc., Paterson, NJ, 1994.

[16] Bandler, J. W., *Computer-Aided Circuit Optimization, in Modern Filter Theory and Design,* G. C. Temes and S. K. Mitra, eds. New York: Wiley, 1973, pp. 211–271.

[17] Bandler, J. W., and S. H. Chen, "Circuit Optimization: The State of the Art," *IEEE Trans. Microwave Theory & Tech.,* Vol. MTT-36, February 1988, pp. 424–443.

[18] *OSA90/hope Reference Manual,* Version 3.5, Optimization System Associates Inc., Canada, 1995.

[19] *Touchstone Reference Manual,* Version 3.0, EEsof Inc., Westlake Village, CA, 1991.

[20] *MDS Reference Manual,* Release 6.0, Hewlett-Packard Co., Palo Alto, CA, 1994.

[21] *Series IV/PC Reference Manual,* Version 6.0, Hewlett-Packard Co., Palo Alto, CA, 1995.

[22] *Super-Compact User's Manual,* Rev. 6.5, Compact Softwaree Inc., Paterson, NJ, 1994.

[23] *LINMIC+ User Manual,* Version 2.1, Jansen Microwave, Germany, 1989.

[24] *M/FILTER Reference Manual,* Eagleware Corp., USA, 1993.

[25] *MMICAD Reference Manual,* Version 2, Optotek Ltd., Canada, 1996.

[26] Cohn, S. B., "Generalized Design of Bandpass and Other Filters by Computer Optimization," *IEEE MTT-S Int. Microwave Symp. Dig.,* June 1974, pp. 272–274.

[27] Postoyalko, V. and D. Budimir, "Design of Waveguide E-Plane Filters With All-Metal Inserts by Equal Ripple Optimization," *IEEE Trans. Microwave Theory & Tech.,* Vol. 42, No. 2, February 1994, pp. 217–222.

[28] Hasler, M., and J. Neiryuck, *Electrical Filters,* Dedham, MA: Artech House, 1986.

[29] Cohn, S. B., "Synthesis of Commensurate Comb-Line Band-Pass Filters With Half-Length Capacitor Lines, and Comparison to Equal-Length and Lumped-Capacitor Cases," *IEEE MTT-S Int. Microwave Symp. Dig.,* May 1980, pp. 135–137.

[30] Hunton, J. K., "Novel Contributions to Microwave Circuit Design," *IEEE MTT-S Int. Microwave Symp. Dig.,* 1989, pp. 753–755.

[31] Budimir, D., and V. Postoyalko, "EPFILTER: A CAD Package for E-Plane Filters," *Microwave Journal,* August 1996, pp. 110–114.

[32] Budimir, D., "Optimized E-Plane Bandpass Filters With Improved Stop Band Performance," *IEEE Trans. Microwave Theory & Tech.,* February 1997, pp. 212–220.

[33] *DBFILTER Reference Manual,* Tesla Communications Ltd., London, England.

[34] Parry, R., "Optimisation of Microwave Filters," *Colloquium on Filters in RF and Microwave Communications,* Digest No. 1992/220, University of Bradford, Bradford, England, December 1992, pp. 7/1–7/5.

[35] *Empipe Reference Manual,* Version 3.1, Optimization System Associates Inc., Canada, 1995.

Selected Bibliography

Bahl, I. J., and P. Bhartia, *Microwave Solid State Circuit Design,* New York: Wiley, 1988.

Empipe3D Reference Manual, Version 3.5, Optimization System Associates Inc., Canada, 1996.

Gupta, K. C., R. Garg, and R. Chadha, *Computer-Aided Design of Microwave Circuits*, Norwood, MA: Artech House, 1981.

Temes, G. C., and D. A. Calahan, "Computer-Aided Network Optimization the State of the Art," *Proc. IEEE*, Vol. 55, 1967, pp. 1832–1863.

2

Transmission Lines, Lumped Elements, and Resonators

This chapter summarizes the information for transmission lines (e.g., coaxial lines, waveguides, striplines, microstrip lines, suspended striplines, *coplanar waveguides* (CPWs), finlines, image guides, microshield lines), lumped elements, and resonators (e.g., dielectric resonator) that is most often needed in filter design. Because of the limited scope of this book, no attempt at completeness has been made. It is hoped that the references included at the end of the chapter will direct the interested reader to sources of more detailed information on particular subjects.

2.1 Transmission Lines

A collection of some transmission line structures commonly used as filter elements in RF, microwave, and mm-wave filters are shown in Figure 2.1. The main applications of these filters can be classified under the following [1]:

- Cellular;
- Cordless;
- *Personal communication systems* (PCSs);
- *Personal communication networks* (PCNs);
- *Wireless local area networks* (WLANs);
- Microwave links;
- Satellite communications;

Circular Coaxial Line

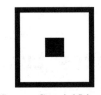

Square Coaxial Line

Rectangular Coaxial Line

Figure 2.1 Transmission line structures commonly used in RF, microwave, and mm-wave filters.

- Radar systems;
- Electronic warfare systems;
- Automotive electronics;
- Microwave instrumentation;
- Troposcatter systems.

The actual choice of transmission line depends on several factors, including the type of filter and its operating frequency. Each type of transmission line has potential advantages for various applications and therefore is briefly discussed here. Frequency band designations (according to radar standard recommendations) are listed in Table 2.1.

2.1.1 Coaxial Line

The basic structure consists of a hollow outer conductor and a center conductor supported by means of a dielectric placed in the space between the outer and

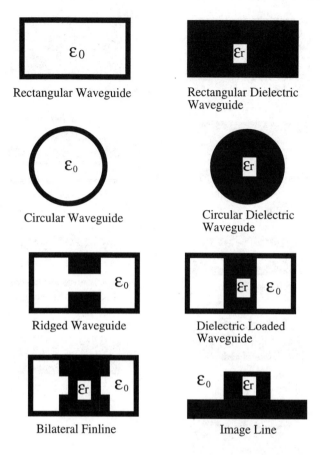

Rectangular Waveguide

Rectangular Dielectric Waveguide

Circular Waveguide

Circular Dielectric Wavegude

Ridged Waveguide

Dielectric Loaded Waveguide

Bilateral Finline

Image Line

Figure 2.1 (continued).

the center conductor. The cross-section of a coaxial line may be circular, square, or rectangular, as shown in Figure 2.1. They often are used in low-power applications at L-band, S-band, C-band, X-band, Ku-band and in some cases as high as Ka-band (see Table 2.1). The dominant mode of propagation is the *transverse electromagnetic* (TEM) mode. The TEM mode has neither an electric nor a magnetic field in the direction of propagation. The mode has no cut-off frequency. The references [2–9] provide a detailed treatment of coaxial lines and include a wide range of basic design details as well as information on discontinuities and applications.

2.1.2 Rectangular Waveguide

Probably the most widely used filter elements in high-power and low-loss microwave and mm-wave filters are rectangular waveguides (see Figure 2.1).

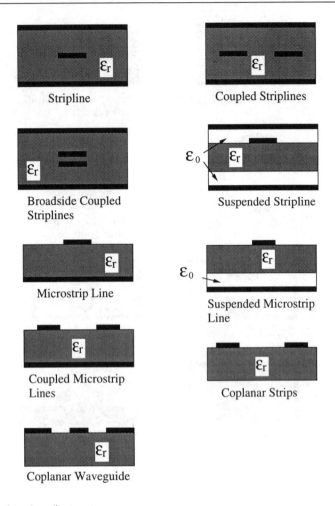

Figure 2.1 (continued).

The waveguide supports *transverse electric* (TE) and *transverse magnetic* (TM) modes. TE modes have a magnetic field (H_z) but no electric field (E_z) in the direction of propagation. They also are referred to as H-modes, or modes of magnetic type.

For TE_{mn} modes, the electromagnetic field components are given by

$$E_x = \frac{jk_0\eta_0 n\pi}{b} A_{mn} \cos\left(\frac{m\pi x}{a}\right) \sin\left(\frac{n\pi y}{b}\right) \qquad (2.1)$$

$$E_y = \frac{jk_0\eta_0 m\pi}{a} A_{mn} \sin\left(\frac{m\pi x}{a}\right) \cos\left(\frac{n\pi y}{b}\right) \qquad (2.2)$$

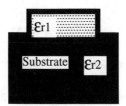

Metal-pipe on Wafer
Waveguide

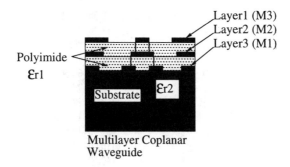

Multilayer Coplanar
Waveguide

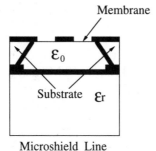

Microshield Line

Figure 2.1 (continued).

$$E_z = 0 \tag{2.3}$$

$$H_x = \frac{j\beta_{mn}m\pi}{a} A_{mn} \quad \sin\left(\frac{m\pi x}{a}\right)\cos\left(\frac{n\pi y}{b}\right) \tag{2.4}$$

$$H_y = \frac{j\beta_{mn}n\pi}{b} A_{mn} \quad \cos\left(\frac{m\pi x}{a}\right)\sin\left(\frac{n\pi y}{b}\right) \tag{2.5}$$

$$H_z = k_c^2 A_{mn} \quad \cos\left(\frac{m\pi x}{a}\right)\sin\left(\frac{n\pi y}{b}\right) \tag{2.6}$$

Table 2.1
Frequency Band Designation

Band Designation	Frequency (GHz)
VHF	0.10–0.30
UHF	0.30–1.00
L	1.00–2.00
S	2.00–4.00
C	4.00–8.00
X	8.00–12.40
Ku	12.40–18.00
K	18.00–26.50
Ka	26.50–40.00
U	40.00–60.50
V	50.00–75.00
E	60.50–92.00
W	75.00–110.00
F	92.00–140.00
D	110.00–170.00
G	140.00–220.00
Y	170.00–260.00
J	220.00–300.00

Modes that have an electric field (E_z) but no magnetic field (H_z) in the direction of propagation are known as TM and also are referred to as E-modes, or modes of electric type. The electromagnetic field components of TM_{mn} modes are given by

$$E_x = \frac{-j\beta_{mn}m\pi}{a} B_{mn} \cos\left(\frac{m\pi x}{a}\right)\sin\left(\frac{n\pi y}{b}\right) \tag{2.7}$$

$$E_y = \frac{-j\beta_{mn}n\pi}{b} B_{mn} \sin\left(\frac{m\pi x}{a}\right)\cos\left(\frac{n\pi y}{b}\right) \tag{2.8}$$

$$E_z = k_c B_{mn} \sin\left(\frac{m\pi x}{a}\right)\sin\left(\frac{n\pi y}{b}\right) \tag{2.9}$$

$$H_x = \frac{jk_0 n\pi}{b\eta_0} B_{mn} \sin\left(\frac{m\pi x}{a}\right)\cos\left(\frac{n\pi y}{b}\right) \tag{2.10}$$

$$H_y = \frac{-jk_0 m\pi}{a\eta_0} B_{mn} \cos\left(\frac{m\pi x}{a}\right)\sin\left(\frac{n\pi y}{b}\right) \tag{2.11}$$

$$H_z = 0 \qquad (2.12)$$

where A_{mn} and B_{mn} are amplitude coefficients, and

$$k_0 = \omega\sqrt{\mu_0\epsilon_0} \qquad (2.13)$$

$$\eta_0 = \sqrt{\frac{\mu_0}{\epsilon_0}} \qquad (2.14)$$

$$k_c = \sqrt{\left(\frac{m\pi}{a}\right)^2 + \left(\frac{n\pi}{b}\right)^2} \qquad (2.15)$$

The modes that enable single-mode operation in the natural state are usually referred to as dominant or fundamental, while all other modes are known as high-order modes. The TE_{10} is the dominant mode of the rectangular waveguide. It has the lowest cutoff frequency of all possible rectangular waveguide modes. Table 2.2 summarizes the properties of standard air-filled rectangular waveguides. Formulas and some design data for them, as well as information on discontinuities and applications, are available in several of the references [2–19]. Because Chapters 3 and 6 of this book are devoted to the study of rectangular waveguide discontinuities and applications, further details are not given here.

2.1.3 Circular Waveguide

A cross-section of a circular waveguide is given in Figure 2.1. Like rectangular waveguides, circular waveguides support TE and TM modes. In the circular waveguide, the dominant mode is the TE_{11}, because it has the lowest cutoff frequency. The circular waveguide is well suited for application in long-distance communication at frequencies between 40 and 110 GHz because it has, if oversized and excited in the TE_{01} mode, extremely low attenuation. However, the TE_{01} is not a dominant mode, thus necessitating the incorporation of mode filters. Additional details on circular waveguides, as well as information on discontinuities and applications, can be found in the references [2–19].

2.1.4 Ridged Waveguide

Ridged waveguides are commonly used in some communication systems for wideband operation. A cross-section of a ridged waveguide is given in Figure 2.1. The introduction of a ridge to the waveguide reduces the cutoff

Table 2.2
Properties of Rectangular Waveguides

EIA WG Designation WR (WG)	Recommended Operating Range for TE_{01} Mode (GHz)	Cutoff Frequency for TE_{01} Mode (GHz)	Theoretical CW Power Rating, Lowest to Highest Frequency (MW)	Inside Dimensions in mm (Inches in parentheses)	
650 (6)	1.12–1.70	0.908	11.90–17.20	165.100–82.550	(6.500–3.250)
510 (7)	1.45–2.20	1.157	7.50–10.70	129.540–64.770	(5.100–2.550)
430 (8)	1.70–2.60	1.372	5.20–7.50	109.220–54.610	(4.300–2.150)
340 (9A)	2.20–3.30	1.736	3.10–4.50	86.360–43.180	(3.400–1.700)
284 (10)	2.60–3.95	2.078	2.20–3.20	72.140–34.040	(2.840–1.340)
229 (11A)	3.30–4.90	2.570	1.60–2.20	58.170–29.083	(2.290–1.145)
187 (12)	3.95–5.85	3.152	1.40–2.00	47.550–22.149	(1.872–0.872)
159 (13)	4.90–7.05	3.711	0.79–1.00	40.390–20.193	(1.590–0.795)
137 (14)	5.85–8.20	4.301	0.56–0.71	34.850–15.799	(1.372–0.622)
112 (15)	7.05–10.00	5.259	0.35–0.46	28.499–12.624	(1.122–0.497)
90 (16)	8.20–12.40	6.557	0.20–0.29	22.860–10.160	(0.900–0.400)
75 (17)	10.00–15.00	7.868	0.17–0.23	19.050–9.525	(0.750–0.375)
62 (18)	12.40–18.00	9.486	0.12–0.16	15.799–7.899	(0.622–0.311)
51 (19)	15.00–22.00	11.574	0.08–0.107	12.954–6.477	(0.510–0.255)
42 (20)	18.00–26.50	14.047	0.043–0.058	10.666–4.318	(0.420–0.170)
34 (21)	22.00–33.00	17.328	0.034–0.048	8.636–4.318	(0.340–0.170)
28 (22)	26.50–40.00	21.081	0.022–0.031	7.112–3.556	(0.280–0.140)
22 (23)	33.00–50.00	26.342	0.014–0.020	5.690–2.845	(0.224–0.112)
19 (24)	40.00–60.00	31.357	0.011–0.015	4.775–2.388	(0.188–0.094)
15 (25)	50.00–75.00	39.863	0.0063–0.0090	3.759–1.880	(0.148–0.074)
12 (26)	60.00–90.00	48.350	0.0042–0.0060	3.099–1.549	(0.122–0.061)
10 (27)	75.00–110.00	59.010	0.0030–0.0041	2.540–1.270	(0.100–0.050)
8 (28)	90.00–140.00	73.840	0.0018–0.0026	2.032–1.016	(0.080–0.040)
7 (136)	110.00–170.00	90.840	0.0012–0.0017	1.651–0.8255	(0.065–0.0325)
5 (135)	40.00–220.00	115.750	0.00071–0.00107	1.2954–0.6477	(0.051–0.0255)
4 (137)	170.00–260.00	137.520	0.00052–0.00075	1.0922–0.5461	(0.043–0.0215)
3 (139)	220.00–325.00	173.280	0.00035–0.00047	0.8636–0.4318	(0.034-0.0170)

frequency of the fundamental mode more than the cutoff of the first higher order mode. Thus, the single-mode operation in such a waveguide can be expanded. By suitable selection of the geometry of the ridge, the bandwidth of the ridged waveguide can be controlled. For more details, the reader is referred to the references [2–9,11,20–34]. Because Chapters 3 and 7 of this book are devoted to the study of ridged waveguides' discontinuities and applications, further details are not given here.

2.1.5 Rectangular Dielectric-Loaded Waveguide

Rectangular dielectric-loaded waveguides are basically a form of capacitively loaded rectangular waveguides. Figure 2.1 shows a cross-sectional view of this type of waveguide, which is used in some communication systems for wideband operation. Like ridged waveguide, the usable bandwidth of dielectric-loaded waveguide is greater than that of an ordinary rectangular waveguide. However it has higher conductor losses and lower power-handling capacity than an ordinary rectangular waveguide. By suitable selection of the geometry of the dielectric slab, the bandwidth of the dielectric-loaded waveguide can be controlled. For more details, the reader is referred to the references [35–37].

2.1.6 E-Plane Circuit (Finline)

A dielectric-loaded waveguide with fins is called a printed E-plane circuit. It can be viewed as a slot line inserted into the E-plane of a rectangular waveguide or as a ridged waveguide with thin ridges backed by dielectric substrate.

It includes all-metal E-plane circuit and finlines (see Figure 2.1). The advantages of low insertion loss, simple fabrication, and wide single-mode bandwidth have made this circuit widely utilized in low- and medium-power mm-wave applications. The mode of propagation is a hybrid mode consisting of a combination of TE and TM modes. In Beyer [38], the line parameters of these waveguides are determined numerically as well as experimentally for the first time considering the metallization thickness and the influence of the longitudinal slits in the mount grove. A substantial amount of work on these structures has been reported in the literature [11,35,38–44]. Because Chapters 3, 6, and 7 of this book are devoted to the study of all-metal E-plane circuits, further details are not given here.

2.1.7 Dielectric Waveguide

At mm-wave frequencies (up to around 140 GHz), where precise fabrication of ordinary rectangular waveguides, suspended striplines, and finlines becomes extremely difficult, dielectric waveguide structures offer an alternative approach with the potential for lower losses and relaxed tolerances. There are several different dielectric waveguide structures, such as dielectric-rod, dielectric-slab (image line), trapped image line, insulated image line, and inverted stripline. Cross-sectional views of dielectric-rod waveguide and image line are shown in Figure 2.1. The dielectric-rod waveguide consists of a rectangular dielectric rod surrounded by an infinite air medium. The dominant mode of a dielectric-rod waveguide has no cutoff frequency, so in theory its operational bandwidth is unlimited. The image line consists of a rectangular dielectric slab placed on a backing sheet of a perfectly conducting ground plane. The structure behaves like a dielectric-rod waveguide, which supports hybrid modes. The main drawback of the structures is the radiation loss at junctions, discontinuities, and bends. Also, the dielectric/metal adhesives are very lossy, which reduces the practical, loaded Q-factor. Because of those difficulties, some further low-cost contenders have been considered for operation at frequencies around and exceeding 100 GHz. A detailed discussion of those structures has been reported in the literature [2,10,11,35–37,45].

2.1.8 Metal-Pipe Waveguide

Monolithic transmission line for submm-wave and terahertz frequency applications may be realized by considering variants of the early dielectric waveguides. The metal-pipe on-wafer waveguides are constructed from dielectric materials and structures, which are available in monolithic technology, so the use of them in integrated circuits is possible. A cross-sectional view of such a transmission line is shown in Figure 2.1. These structures are used not only as transmission lines but also as filter elements [46]. For more informations about these lines, the reader is referred to the references [46–49].

2.1.9 Stripline

One of the most commonly used homogeneous planar transmission lines in a microwave low-power transmission system is stripline. A cross-sectional view of stripline is given in Figure 2.1. The dominant mode of propagation in this type of line is the TEM mode. The basic structure consists of the flat strip conductor situated symmetrically between two large ground planes. The electric and magnetic fields are concentrated around the strip conductor and away from the strip fields that decay rapidly with distance. Formulas and some design data for these lines and applications are available in the literature [6-9,11,50–55].

Suspended and coupled striplines are the most useful variants of stripline. Basically, the suspended stripline (see Figure 2.1) is an inhomogeneous line in which the substrate carrying the strip conductor is placed symmetrically between the two ground planes, thereby leaving an air gap on either side of the substrate. Reduced thickness of the substrate partly decreases the dielectric losses. However, the considerable conduction losses, particularly at the edges of the strips, remain unchanged. The shielding eliminates transverse coupling and radiation but contributes to additional costs. The possibility of breakage of the thin substrate is a disadvantage that cannot be disregarded. The configuration of parallel coupled stripline is shown in Fig 2.1. The principal application areas of parallel coupled striplines are filters and a variety of other useful circuits.

2.1.10 Microstrip Line

Microstrip is the second basic type of planar transmission line, developed immediately after stripline. Unlike stripline, microstrip is an inhomogeneous structure with quasi-TEM as the dominant mode of propagation. A variant of this line, suspended microstrip, incorporates an air gap between the substrate and the ground (see Figure 2.1). It provides a higher Q (500 to 1,500) than microstrip. The wide range of impedance values achievable makes these lines particularly suitable for filters.

A coupled microstrip line configuration consists of two transmission lines placed parallel to each other and in close proximity. Coupled lines are utilized extensively as basic elements for filters and a variety of other useful circuits. Because of the coupling of electromagnetic fields, a pair of coupled lines can support two different modes of propagation. These modes have different characteristic impedances. When the lines are imbedded in a homogeneous dielectric medium the velocity of propagation of these two modes is equal. However, for transmission lines such as coupled microstrip lines the dielectric medium is not homogeneous. In the configuration for the parallel coupled microstrip lines, a part of the field extends into the air above the substrate (see Figure 2.1). This fraction of total field is different for the two modes of the coupled lines. Consequently, the effective dielectric constant and the phase velocities are not equal for the two modes. This nonsynchronous feature reduces the performance of circuits using these types of coupled lines. Numerous papers and a few books [3,7,8,10,11,44,50,56–74] deal with the analysis of single, suspended, and coupled microstrip lines, characterization of discontinuities, and various other aspects such as losses, dispersive behavior, and applications.

2.1.11 Coplanar Line

Coplanar waveguide and coplanar strip structures (see Figure 2.1) belong to the category of coplanar lines. All conductors are on the same side of a dielectric substrate. The dominant mode of propagation in the coplanar lines is a quasi-TEM mode, because of different dielectrics above and below the conductors. A CPW consists of a strip conductor placed on a dielectric substrate with two ground planes adjacent and parallel to the strip, while a coplanar strip structure comprising a pair of coplanar strips is placed on a dielectric substrate. Coplanar strips are used more in high-speed digital circuits then in microwave circuits. A detailed discussion of these lines, characterization of discontinuities, and various other aspects such as losses, dispersive behavior, and applications can be found in the references [3,7,8,10,50,56,57,72,74].

2.1.12   Multilayer Coplanar Waveguide

With dielectric and conductor layers, additional degrees of freedom arise for filter design, like an extended range of impedances. In the case of *Monolithic Microwave Integrated Circuit* (MMIC) filters, the additional layer can be added using a thin film of polyimide or silicon nitride. For example, a fifth-order lowpass CPW multilayer MMIC filter was tested (Figure 2.2). It consists of three low-impedance V-shaped multilayer and high-impedance elevated CPW transmission lines on a GaAs substrate. With reference to Figure 2.1, the top, middle, and bottom sections of the structure are made from aluminum metallization layers, M3, M2, and M1, respectively. Layers of polyimide are employed for separation of the two metal layers M3/M2 and M2/M1. A photograph of a V-shaped multilayer CPW is shown in Figure 2.3. A description of this transmission structure is available in the literature [75–81].

2.1.13   Microshield Line

Microshield line was introduced as an alternative transmission medium to CPW for mm-wave and submm-wave applications. Dib et al. describes this

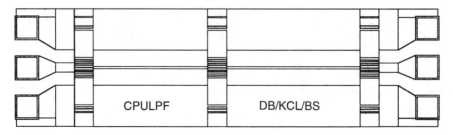

Figure 2.2 Layout of lowpass multilayer coplanar waveguide filter.

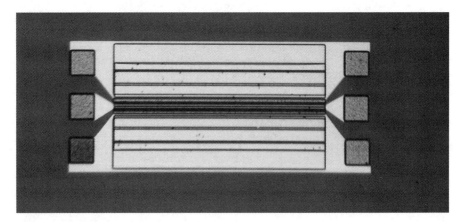

Figure 2.3 Photomicrograph of multilayer coplanar waveguide.

type of waveguide extensively [83]. It is a half-shielded geometry, quasi-planar transmission line similar to CPW, which uses a thin (around 1.5 μm) dielectric membrane to support the conducting lines and upper ground planes above a metallized shielding cavity. That configuration allows single-mode TEM wave propagation over a broad bandwidth with zero dielectric loss and minimal dispersion. A description of these and other advantages of the microshield geometry can be found in the literature [82–86].

2.2 Comparison of Various Transmission Lines

The preceding sections briefly described several different types of transmission line structures. For a given frequency range, the selection of a particular transmission line structure is based on several considerations, some of which are presented in Table 2.3.

2.3 Lumped Elements

This section briefly describes commonly used lumped elements (as filter elements). Detailed discussion of lumped elements has been reported in the literature [7,8,56,87–91]. Filter elements can be both lumped elements (dimensions <0.1 wavelength) and distribution elements that are composed of sections of transmission lines and waveguides. The choice of filter elements depends on the frequency of operation. At lower frequencies, the lumped element filters that have a lower Q than distributed element filters have the advantage of lower cost, smaller size, and wideband performances. However, it is often

Table 2.3
Impedance Ranges, Unloaded Q-Factors, and Frequency Ranges for Various Transmission Line Structures

Transmission Line Structure	Characteristic Impedance Range (Ω)	Unloaded Q-Factor (Q_u) at 30 GHz	Frequency Range (GHz)
Microstrip	20–125	250	up to 110
CPW	25–150	100	up to 100
Suspended stripline	40–150	up to 600	up to 110
Suspended microstrip	40–150	500–1,500	up to 110
Finline	10–400	500	up to 110
Image line	$\approx$26	2,700	up to 110

difficult to realize a truly lumped element even at lower frequencies because of the parasitics to ground associated with thin substrates. With the advent of new techniques [91], the fabrication of lumped elements can now be extended from X-band to about 100 GHz. The two basic building blocks for filter design available in lumped form are inductors and capacitors. The design of lumped element filters by computer optimization (described in Chapter 5) requires a complete and accurate characterization of lumped elements at high frequencies. That necessitates the development of electromagnetic models, which take into account the presence of ground planes, proximity effect, fringing fields, parasitics, and so on.

2.3.1 Inductors

Depending on the inductance required and the frequency of operation, inductors can be realized in a few types and all are achieved with a single-layer metallization scheme. High-impedance line section, meander inductor, and square spiral inductor are illustrated in Figure 2.4. Table 2.4 compares various inductors.

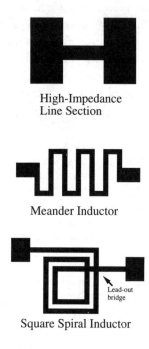

High-Impedance
Line Section

Meander Inductor

Lead-out
bridge

Square Spiral Inductor

Figure 2.4 Planar inductor configurations.

Table 2.4
Comparison of Various Planar Inductors

Inductor Type	Inductance Range (nH)	Unloaded Q Factor (Q_u) at 10 GHz
Straight section of line	2–3	50
Spiral (circular, square, or rectangular)	>3	>50

2.3.2 Capacitors

A lumped capacitor can be realized by the use of a single metallization scheme and a two-level metallization technology in conjunction with dielectric films. Figure 2.5 shows a low-impedance transmission line, end-coupled, and interdigital capacitor. Choosing from these capacitors depends on the capacitance required, the frequency of operation, and the processing technology available. Table 2.5 compares three types of capacitors. Figure 2.2 illustrated a lowpass filter realized by cascaded sections of high-impedance multilayer CPWs (inductors) and low-impedance multilayer CPWs (capacitors) (see Figure 2.1).

2.4 Resonators

Structures extensively used as filter elements in the realization of various bandpass and bandstop filters are resonant structures. At low frequencies, resonant

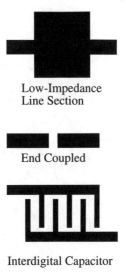

Low-Impedance
Line Section

End Coupled

Interdigital Capacitor

Figure 2.5 Planar capacitor configurations.

Table 2.5
Comparison of Various Planar Capacitors

Capacitor Type	Capacitance Range (pF)	Unloaded Q Factor (Q_u) at 10 GHz
Straight section of line	<0.2	50
End coupled	<0.2	50
Interdigital	0.05–0.5	50
Metal insulator metal (overlay)	0.1–30	50–100

structures invariably are composed of lumped elements. At higher frequencies, however, lumped elements in general cannot be employed because of low Q and difficulties in realizing inductances and capacitances at high frequencies. Distributed elements are widely used to overcome those limitations. The resonant structures commonly used in various bandpass and bandstop filters can be realized by several forms, such as lumped element resonators, cavity resonators (Figure 2.6), planar resonators (Figure 2.7), and dielectric resonators (Figure 2.8). Choosing from these forms depends on the mechanical size of the resonator at the resonant frequency, the unloaded Q, and the temperature stability and electrical tunability of the frequency. This section presents a brief description of the dielectric resonators commonly used in various bandpass and bandstop filters. A detailed description of resonant structures can be found in the literature [2–4,8,56,92–98].

2.4.1 Dielectric Resonator

A dielectric resonator consists of a solid, low-loss, temperature-stable, high-permitivity ($100 > \epsilon_r > 30$), and high-Q dielectric material, formed in a few geometrical shapes, as shown in Figure 2.8. It resonates in various modes at

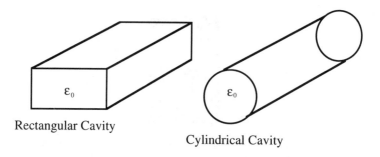

Rectangular Cavity

Cylindrical Cavity

Figure 2.6 Cavity resonators.

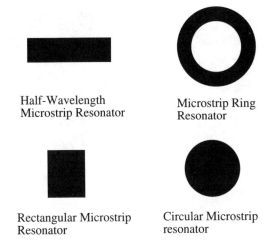

Figure 2.7 Planar resonators.

frequencies determined by resonator dimensions and shielding conditions. Typical geometrical shapes used for dielectric resonators are as follows:

- Disc type;
- Cylinder type;
- Dielectric-loaded type;
- TEM coaxial dielectric resonator;
- Distributed Bragg reflector resonator.

With their high unloaded Q (Q_u), high dielectric constant, and high-temperature-stability dielectric resonators are used in place of waveguide components in a large variety of microwave systems to reduce their size by a factor of approximately $1/\sqrt{\epsilon_r}$ for equal electrical performances. The intrinsic loss of a dielectric material can be expressed in terms of its loss tangent (tan δ). If the RF field is totally contained within the dielectric, then the resonator unloaded Q is given as

$$Q_U = \frac{1}{\tan \delta} = Q_D \qquad (2.16)$$

If the field extends outside the dielectric (as it will in practice), it will excite currents and produce ohmic loss in the surrounding enclosure. In that case,

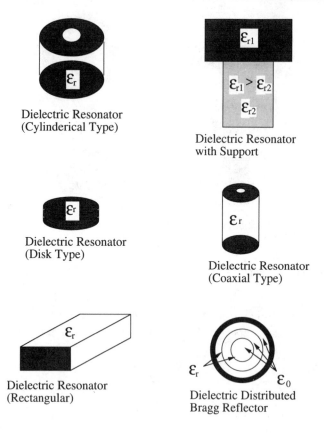

Dielectric Resonator
(Cylinderical Type)

Dielectric Resonator
with Support

Dielectric Resonator
(Disk Type)

Dielectric Resonator
(Coaxial Type)

Dielectric Resonator
(Rectangular)

Dielectric Distributed
Bragg Reflector

Figure 2.8 Dielectric resonators.

$$\frac{1}{Q_U} = \frac{1}{Q_D} + \frac{1}{Q_C} \tag{2.17}$$

where Q_c is the Q factor due to conductor losses only. It is apparent that the high-resonator Q possible with low-loss dielectrics is achievable only if Q_c can be maximized. In practical cases, the enclosure losses may dominate the resonator Q or determine the maximum amount of miniaturization that can be achieved for a given Q_u. The actual loss of a dielectric resonator will be determined by the following:

- Loss tangent of dielectric material;
- Mode of operation, which is the field configuration at resonance and is determined by the geometry and mounting of the dielectric on its enclosure;

- Loss tangent of support material: many modes of operation require the dielectric resonator to be positioned in the center of a cavity, which is achieved in practice by positioning the puck on a low dielectric constant support;
- RF loss of adhesive used to secure dielectric;
- Resistivity and size of metallic enclosure;
- Tuning and coupling structures.

The resonator bandwidth is inversely proportional to the Q factor; thus, high Q-factor resonators have narrow bandwidths.

2.4.2 Dielectric Material

The important properties of the dielectric material used in dielectric resonators are as follows:

- Q factor, which is equal to the inverse of the loss tangent. The Q factor of the dielectric resonator decreases with increasing in frequency.
- Dielectric constant.
- The temperature coefficient at the resonant frequency, T_f, which includes the combined effects of the temperature coefficient of the dielectric constant and the thermal expression of the dielectric. The temperature coefficient, T_f, of the resonator can be controlled in some materials by modifying the composition.

Manufacturers' data from a selection of available materials are given in Tables 2.6–2.10. The loss is given as the dielectric Q, which is the reciprocal

Table 2.6
Properties of Dielectric Materials Manufactured by NTK Piezoelectric Ceramics [97]

	ϵ_r	Q	Operating Frequency (GHz)
A material	12.6	18,000	13.0
C material	21.0	12,000	6.0
F material	34.0	8,000	8.0
E material	45.0	10,600	4.0
T material	80.0	1,500	2.7
L material	93.0	1,350	2.7
S material	6.40	2,200	10.0

Table 2.7
Properties of Dielectric Materials Manufactured by Tekelec Components [98]

	ϵ_r	Q	Operating Frequency (GHz)
E 2000 series	36.9–37.2	9,500	5.0
E 9000 series	36.0–37.0	40,000	1.0

Table 2.8
Properties of Some Typical Dielectric Materials [99]

	ϵ_r	Q	Operating Frequency (GHz)
Sapphire	9.394	450,000	13.2
Sapphire	9.394	650,000	9.0

of the loss tangent. The loss tangent normally is linear with frequency, so that for a given material the product of Q and frequency is a constant. That enables the loss to be extrapolated to other frequencies. The main purpose of Tables 2.6–2.10 is to give an idea of what is typically known about the properties of materials used in dielectric resonators.

Table 2.9
Properties of Dielectric Materials Manufactured by Alpha/Trans-Tech, Inc. [100]

8300 Series ϵ_r = 35–36.5 Q > 28,000 at 850 MHz
Composition: Barium titanate
Frequency range: Disc type: 800–13,800 MHz
Frequency range: Cylinder type (ring): 800–9010 MHz
Temperature coefficient of $TE_{01\delta}$ mode resonant frequency, T_f (ppm/C) $\pm$ (2.0 or 1.0)

8600 Series ϵ_r = 80.0 Q > 3,000 at 3.000 GHz
Composition: BaLnTi Oxide
Frequency range: Disc type: 700–3618 MHz
Frequency range: Cylinder type (ring): 967–3618 MHz
Temperature coefficient of $TE_{01\delta}$ mode resonant frequency, T_f (ppm/C) $\pm$ (2.0 or 1.0)

8700 Series ϵ_r = 27.6–30.6 Q > 10,000 at 10.0 GHz
Composition: BaZnTaTi Oxide
Frequency range: Disc type: 5550–32150 MHz
Frequency range: Cylinder type (ring): 5550–9870 MHz
Temperature coefficient of $TE_{01\delta}$ mode resonant frequency, T_f (ppm/C) $\pm$ (2.0 or 1.0)

8800 Series ϵ_r = 36.6–38.3 Q > 6,000 at 4.5 GHz
Composition: BaTitanium Oxide
Frequency range: Disc type: 790–5210 MHz
Frequency range: Cylinder type (ring): 1390–5210 MHz
Temperature coefficient of $TE_{01\delta}$ mode resonant frequency, T_f (ppm/C) $\pm$ (2.0 or 1.0)

9000 Series ϵ_r = 85.5 Q > 1,500 at 1.0 GHz
Frequency range: TEM mode resonator of rod: 320–5000 MHz

Table 2.10
Properties of Some Typical Dielectric Materials Manufactured by Murata Manufacturing
Co., Ltd. [101]

Brand: Resomics-U series	ϵ_r = 36.0–40.0	Q = 6,000 at 7.0 GHz
Composition: (Zr,Sn) TiO4		
Frequency range: 1–12 GHz		
Brand: Resomics-R series	ϵ_r = 36.0–40.0	Q = 6,000 at 7.0 GHz
Composition: (Zr,Sn) TiO4		
Frequency range: 1–12 GHz		
Brand: Resomics-R series	ϵ_r = 36.0–40.0	Q = 6,000 at 7.0 GHz
Composition: (Zr,Sn) TiO4		
Frequency range: 1–12 GHz		
Brand: Resomics-E series	ϵ_r = 36.0–40.0	Q = 6,000 at 7.0 GHz
Composition: (Zr,Sn) TiO4		
Frequency range: 1–12 GHz		
Brand: Resomics-F series (Disc type)	ϵ_r = 24	Q = 35,000 at 10 GHz
Frequency range: 10–25 GHz		
Brand: Resomics-F series (Ring type)	ϵ_r = 24	Q = 35,000 at 10 GHz
Frequency range: 10–19.5 GHz		

Brand: Resomics-TEM mode resonator of rod (silver-plated) material:
U ϵ_r = 37–39 Q = 300 at 2500 MHz
K ϵ_r = 91–93 Q = 500 at 1500 MHz
Frequency range: 400–4800 MHz

Brand: Resomics-TEM mode resonator of rod (copper-plated) material:
K ϵ_r = 91–93 1–5 Q = 500 at 1500 MHz
Frequency range: 500–3000 MHz

References

[1] Special Issue on Emerging Commercial and Consumer Circuits, Systems, and Their Applications, *IEEE Trans. Microwave Theory & Tech.,* Vol. MTT-43, July 1995.

[2] Ramo, S., J. R. Whinnery, and T. Van Duzer, *Fields and Waves in Communication Electronics,* 3rd ed., New York: Wiley, 1994.

[3] Collin, R. E., *Foundations for Microwave Engineering,* 2nd ed., New York: McGraw-Hill, 1992.

[4] Rizzi, P. A., *Microwave Engineering Passive Circuits,* Englewood Cliffs, NJ: Prentice-Hall, 1988.

[5] Pozar, D. M., *Microwave Engineering,* Reading, MA: Addison-Wesley, 1990.

[6] Matthaei, G., L. Young, and E. M. T. Yones, *Microwave Filters, Impedance Matching Networks, and Coupling Structures*, 2nd ed., Norwood, MA: Artech House, 1980.

[7] Gupta, K. C., R. Garg, and R. Chadha, *Computer-Aided Design of Microwave Circuits*, Norwood, MA: Artech House, 1981.

[8] Bahl, I. J., and P. Bhartia, *Microwave Solid State Circuit Design*, New York: Wiley, 1988.

[9] Saad, T. S., *Microwave Engineers Handbook, Vol. 1*, Dedham, MA: Artech House, 1971.

[10] Bhartia, P., and I. J. Bahl, *Millimeter Wave Engineering and Applications*, New York: Wiley, 1984.

[11] Chang, K., ed., *Handbook of Microwave and Optical Components, Vol. 1*, New York: Wiley, 1989.

[12] Marcuvitz, N., *Waveguide Handbook*, Rad. Lab. Series, Vol. 10, New York: McGraw-Hill, 1951.

[13] Schwinger, J., and D. Saxon, *Discontinuities in Wave Guide (Documents on Modern Physics)*, New York: Gordon and Greach, 1968.

[14] Collin, R. E., *Field Theory of Guided Waves*, 2nd ed., Piscataway, NJ: IEEE Press, 1990.

[15] Wexler, A., "Solution of Waveguide Discontinuities by Modal Analysis," *IEEE Trans. Microwave Theory & Tech.*, Vol. MTT-15, September 1967, pp. 508–517.

[16] Kuhn, E., "A Mode-Matching Method for Solving Field Problems in Waveguide and Resonator Circuits," *AEU*, Band 27, Heft 12, 1973, pp. 511–518.

[17] Patzelt, H., and F. Arndt, "Double-Plane Steps in Rectangular Waveguide and Their Application for Transformers, Irises and Filters," *IEEE Trans. Microwave Theory & Tech.*, Vol. MTT-30, May 1982, pp. 771–776.

[18] Shih, Y. C., and K. Gray, "Convergence of Numerical Solutions of Step-Type Waveguide Discontinuity Problems by Modal Analysis," *IEEE MTT-S Int. Microwave Symp. Dig.*, 1983, pp. 233–235.

[19] Tao, J. W., and H. Baudrand, "Multimodal Variational Analysis of Uniaxial Waveguide Discontinuities," *IEEE Trans. Microwave Theory & Tech.*, Vol. MTT-39, March 1991, pp. 506–516.

[20] Cohn, S. B., "Properties of Ridge Waveguide," *Proc. IRE*, Vol. 35, August 1947, pp. 783–788.

[21] Hopfer, S., "The Design of Ridged Waveguide," *IRE Trans. Microwave Theory & Tech.*, Vol. MTT-3, October 1955, pp. 20–29.

[22] Pyle, J. R., "The Cutoff Wavelength of the TE10 Mode in Ridged Rectangular Waveguide of Any Aspect Ratio," *IEEE Trans. Microwave Theory & Tech.*, Vol. MTT-14, April 1966, pp. 175–183.

[23] Montgomery, J. P., "On the Complete Eigenvalue Solution or Ridged Waveguide," *IEEE Trans. Microwave Theory & Tech.*, Vol. MTT-19, June 1971, pp. 547–555.

[24] Dasgupta, D., and P. K. Saha, "Eigenvalue Spectrum of Rectangular Waveguide With Two Symmetrically Placed Double Ridges," *IEEE Trans. Microwave Theory & Tech.*, Vol. MTT-29, January 1981, pp. 47–51.

[25] Hoefer, W. J. R., and M. R. Burtin, "Closed-Form Expressions for the Parameters of Finned and Ridged Waveguide," *IEEE Trans. Microwave Theory & Tech.*, Vol. MTT-30, December 1982, pp. 2190–2194.

[26] Utsumi, Y., "Variation Analysis of Ridged Waveguide Modes," *IEEE Trans. Microwave Theory & Tech.*, Vol. MTT-33, February 1985, pp. 111–120.

[27] Tao, J. W., and H. Baudrand, "Rigorous Analysis of Triple-Ridge Waveguides," *Electronics Letters*, Vol. 24, No. 13, June 1988, pp. 820–821.

[28] Bornemann, J., and F. Arndt, "Transverse Resonance, Standing Wave and Resonator Formulations of the Ridge Waveguide Eigenvalue Problem and Its Application to the Design of E-Plane Finned Waveguide Filters," *IEEE Trans. Microwave Theory & Tech.*, Vol. MTT-38, August 1990, pp. 1104–1113.

[29] Bornemann, J., "Comparison Between Different Formulations of the Transverse Resonance Field-Matching Technique for the Three Dimensional Analysis of Metal-Fined Waveguide Resonators," *Int. J. Numerical Modelling*, Vol. 4 , March 1991, pp. 63–73.

[30] Fan, P., and D. Fan, "Computer Aided Design of E-Plane Waveguide Passive Components," *IEEE Trans. Microwave Theory & Tech.*, Vol. MTT-37, February 1989, pp. 335–339.

[31] Gebauer, A., and F. Hernandez-Gil, "Analysis and Design of Waveguide Multiplexers Using the Finite Element Method," *Proc. 18th European Microwave Conf.*, Stockholm, Sweden, 1988, pp. 521–524.

[32] Omar, A. S., and K. Schunemann, "Application of the Generalized Spectral-Domain Technique to the Analysis of Rectangular Waveguides With Rectangular and Circular Metal Inserts," *IEEE Trans. Microwave Theory & Tech.*, Vol. MTT-39, June 1991, pp. 944–952.

[33] Getsinger, W. J., "Ridged Waveguide Field Description and Application to Direction Couplers," *IRE Trans. Microwave Theory & Tech.*, Vol. MTT-10, January 1962, pp. 41–51.

[34] Konishi, Y., and H. Matsumura, "Short End Effect of Ridge Guide With Planar Circuit Mounted in Waveguide," *IEEE Trans. Microwave Theory & Tech.*, Vol. MTT-26, October 1978, pp. 716–719.

[35] Mansur, R. R., R. S. K. Tong, and R. H. Macphie, "Simplified Description of the Field Distribution in Finlines and Ridge Waveguide and Its Application to the Analysis of E-Plane Discontinuities," *IEEE Trans. Microwave Theory & Tech.*, Vol. MTT-36, December 1988, pp. 1825–1832.

[36] Knox, R., "Dielectric Waveguide Microwave Integrated Circuits—An Overview," *IEEE Trans. Microwave Theory & Tech.*, Vol. MTT-24, No. 11, November 1976, pp. 806–814.

[37] Solbach, K., "The Fabrication of Dielectric Image Lines Using Casting Resins and the Properties of the Lines in the Millimeter Wave Range," *IEEE Trans. Microwave Theory & Tech.*, Vol. MTT-24, November 1976, pp. 879–881.

[38] Beyer, A., "Analysis of the Characteristics on an Earthed Fin Line," *IEEE Trans. Microwave Theory & Tech.*, Vol. MTT-29, July 1981, pp. 676–680.

[39] Meier, P., "Integrated Fin-Line Millimeter Components," *IEEE Trans. Microwave Theory & Tech.*, Vol. MTT- 22, No. 12, December 1974, pp. 1209–1216.

[40] Solbach, K., "The Status of Printed Millimeter Wave E-Plane Circuits," *IEEE Trans. Microwave Theory & Tech.*, Vol. MTT-31, February 1983, pp. 107–121.

[41] Bhat, B., and S. Koul, "Analysis Design and Applications of Finlines" Norwood, MA: Artech House, 1987.

[42] Omar, A. S., and K. Schunemann, "Transmission Matrix Representation of Finline Discontinuities," *IEEE Trans. Microwave Theory & Tech.,* Vol. MTT-33, September 1985, pp. 765–770.

[43] Bornemann, J., and F. Arndt, "Modal S-Matrix Design of Optimum Stepped Ridged and Finned Waveguide Transformers," *IEEE Trans. Microwave Theory & Tech.,* Vol. MTT-35, June 1987, pp. 561–567.

[44] Gupta, K. C., et al., *Microstrip Lines and Slotlines,* 2nd ed., Norwood, MA: Artech House, 1996.

[45] Gelsthorpe, R. V., N. Williams, and N. M. Davey, "Dielectric Waveguide: A Low Cost Technology for Millimeter Wave Integrated Circuits," *Radio and Electronics Engineer,* Vol. 52, November-December 1982, pp. 522–528.

[46] Brown, D. A., A. S. Treen, and N. J. Cronin, "Micromachining of Terahertz Waveguide Components With Integrated Active Devices," *19th Int. Conf. Infrared and Millimetre-waves,* Sendai, Japan, 1994.

[47] Yeh, C., F. I. Shimabukuro, and J. Chu, "Dielectric Ribbon Waveguides: An Optimum Configuration for Ultra-Low-Loss Millimetre/Submillimeter Dielectric Waveguide," *IEEE Trans. Microwave Theory & Tech.,* Vol. MTT-38, June 1990.

[48] Engel, A. G., Jr., and L. P. B. Katehi, "Low-Loss Monolithic Transmission Lines for Submillimeter and Teraherz Frequency Applications," *IEEE Trans. Microwave Theory & Tech.,* Vol. MTT-39, November 1991, pp. 1847–1854.

[49] Treen, A. S., and N. J. Cronin, "Terahertz Metal-Pipe Waveguides," *18th Int. Conf. Infrared and Millimetre-Waves,* Essex, UK, 1993, pp. 470–471.

[50] Hoffman, R. K., *Handbook of Microwave Integrated Circuits,* Norwood, MA: Artech House, 1987.

[51] Lucyszyn, S., et al., "Design of Compact Monolithic Dielectric Filled Metal-Pipe Rectangular Waveguide for Millimetre-Wave Applications," *IEE Proc. Microwave, Antennas and Propagation,* Part H, Vol. 143, No. 5, October 1996.

[52] Bahl, I. J., and R. Garg, "Designer's Guide to Stripline Circuits," *Microwaves,* Vol. 17, January 1978, pp. 90–96.

[53] Cohn, S. B., "Characteristic Impedance of Shielded Strip Transmission Line," *IRE Trans. Microwave Theory Tech.,* Vol. MTT-2, July 1954, pp. 52–55.

[54] Howe, H., Jr., *Stripline Circuits Design,* Dedham, MA: Artech House, 1974.

[55] Barrett, R. M., "Microwave Printed Circuits—The Early Years," *IEEE Trans. Microwave Theory & Tech.,* Vol. MTT-32, No. 9, September 1984, pp. 983–990.

[56] Konishi, Y., ed., *Microwave Integrated Circuits,* New York: Dekker, 1991.

[57] Howe, H., Jr., "Microwave Integrated Circuits-An Historical Perspective," *IEEE Trans. Microwave Theory & Tech.,* Vol. MTT-32, No. 9, September 1984, pp. 991–996.

[58] Edwards, T., *Foundation for Microstrip Circuit Design,* 2nd ed., Chichester, England: Wiley, 1992.

[59] Gardiol, F., *Microstrip Circuit,* New York: Wiley, 1994.

[60] Frey, J., and K. Bhasin, eds., *Microwave Integrated Circuits,* Norwood, MA: Artech House, 1985.

[61] Schneider, M. V., "Microstrip Lines for Microwave Integrated Circuits," *Bell Syst. Tech. J.,* Vol. 48, May-June 1969, pp. 1421–1444.

[62] Denlinger, E. J., "A Frequency Dependent Solution for Microstrip Transmission Lines," *IEEE Trans. Microwave Theory & Tech.*, Vol. MTT-19, No. 1, Jan. 1971, pp. 30–39.

[63] Bryant, T. G., and J. A. Weiss, "Parameters of Microstrip Transmission Lines and of Coupled Pairs of Microstrip Lines," *IEEE Trans. Microwave Theory & Tech.*, Vol. MTT-16, No. 12, December. 1968, pp. 1021–1027.

[64] Silvester, P., "TEM Wave Properties of Microstrip Transmission Lines," *Proc. IEEE*, Vol. 115, 1968, pp. 43–48.

[65] Schneider, M. V., "Microstrip Lines for Microwave Integrated Circuits," *Bell System Tech. J.*, Vol. 48, 1969, pp. 1421–1444.

[66] Garg, R., "Design Equations for Coupled Microstrip Lines," *Int. J. Electron.*, Vol. 47, 1979, pp. 587–591.

[67] Garg, R., and I. J. Bahl,, "Characteristics of Coupled Microstriplines," *IEEE Trans. Microwave Theory & Tech.*, Vol. MTT-27, July 1979, pp. 700–705.

[68] Getsinger, W. J., "Dispersion of Parallel-Coupled Microstrips," *IEEE Trans. Microwave Theory & Tech.*, Vol. MTT-21, 1973, pp. 144–145.

[69] Akhtarzad, S., et al, "The Design of Coupled Microstrip Lines," *IEEE Trans. Microwave Theory & Tech.*, Vol. MTT-23, June 1975, pp. 486–492.

[70] Hinton, J. H., "On Design of Coupled Microstrip Lines," *IEEE Trans. Microwave Theory & Tech.*, Vol. MTT- 28, March 1980, pp. 272.

[71] Vendelin, G. D., "Limitations on Stripline Q," *Microwave J.*, Vol. 13, May 1970, pp. 63–69.

[72] Mittra, R., and T. Itoh, "Analysis of Microstrip Transmission Lines," in *Advances in Microwaves*, Vol. 8, New York: Academic Press, 1974.

[73] Itoh, T., "Overview of Quasi-Planar Transmission Lines," *IEEE Trans. Microwave Theory & Tech.*, Vol. MTT-37, February 1989, pp. 275–280.

[74] Chu, T. S., and T. Itoh, "Comparative Study of Mode-Matching Formulations for Microstrip Discontinuity Problems," *IEEE Trans. Microwave Theory & Tech.*, Vol. MTT-33, October 1985, pp. 1018–1023.

[75] Wen, C. P., "Coplanar Waveguide: A Surface Strip Transmission Line Suitable for Nonreciprocal Gyromagnetic Device Application," *IEEE Trans. Microwave Theory & Tech.*, Vol. MTT-17, No. 12, December 1969, pp. 1087–1090.

[76] Budimir, D., et al., "V-Shaped CPW Transmission Lines for Multilayer MMICs," *IEE Electronics Letters*, October 1995, pp. 1928–1929.

[77] Wang, Q., et al., "Fabrication and Microwave Characterisation of Multilayer Circuits for MMIC Applications," *IEE Proceedings*, Part H, June 1996.

[78] Budimir, D., et al., "Low Loss Multilayer Coplanar Waveguide Transmission Lines on Silicon Substrate for MMICs," *Proc. 26th European Microwave Conf.*, Prague, Czech Republic, September 1996.

[79] Budimir, D., et al., "Novel Transmission-Lines With V-shaped Conductors for HTS Circuit and MMIC Applications," *IEE Proc. Microwave, Antennas and Propagation*, Part H, 1997.

[80] Budimir, D., et al., "CPW Lowpass Filters on Silicon Substrate for Multilayer MMICs," *Microwave and Optical Technology Letters*, 1997.

[81] Gokdemir, T., et al., "Multilayer Passive Components for Uniplanar Si/SiGe MMICs," *IEEE MTT-S Int. Microwave Symp. Digest,* June 1997.

[82] Dib, N. I., et al., "Study of a Novel Planar Transmission Line," *IEEE MTT-S Int. Microwave Symp. Digest,* 1991, pp. 623-626.

[83] Robertson, S. V., L. P. B. Katehi, and G. M. Rebeiz, "W-Band Microshield Low-Pass Filters," *IEEE MTT-S Int. Microwave Symp. Digest,* 1994, pp. 625–628.

[84] Weller, T. M., and L. P. B. Katehi, "Miniature Stub and Filter Designs Using the Microshield Transmission Line," *IEEE MTT-S Int. Microwave Symp. Digest,* 1995, pp. 675–678.

[85] Weller, T. M., L. P. B. Katehi, and G. M. Rebeiz, "High Performance Microshield Line Components," *IEEE Trans. Microwave Theory & Tech.,* Vol. MTT-43, March 1995, pp. 534–543.

[86] Weller, T. M., L. P. B. Katehi, and G. M. Rebeiz, "Novel Micromachined Approaches to MMICs Using Low-Parasitic, High-Performance Transmission Media and Environments," *IEEE MTT-S Int. Microwave Symp. Digest,* 1996, pp. 1145–1148.

[87] Caulton, M., "Lumped Elements in Microwave Integrated Circuits," in *Advances in Microwaves,* Vol. 8, L. Young and H. Sobol, eds., New York: Academic Press, 1974.

[88] Aitchison, C. S., et al., "Lumped-Circuit Elements at Microwave Frequencies," *IEEE Trans. Microwave Theory & Tech.,* Vol. MTT-19, December 1971, pp. 928–937.

[89] Daly, D. A., et al., "Lumped Elements in Microwave Integrated Circuits," *IEEE Trans. Microwave Theory & Tech.,* Vol. MTT-15, December 1967, pp. 713–721.

[90] Pucel, R. Q., "Design Consideration for Monolithic Microwave Circuits," *IEEE Trans. Microwave Theory & Tech.,* Vol. MTT-29, No. 6, June 1981, pp. 513–534.

[91] Chi, C-Y., and G. Rebeiz, "Planar Microwave and Millimeter-Wave Lumped Elements and Coupled-Line Filters Using Micro-Machining Techniques," *IEEE Trans. Microwave Theory & Tech.,* Vol. MTT-43, April 1995, pp. 730–738.

[92] Kajfez, D., and P. Guillon, *Dielectric Resonators,* Norwood, MA: Artech House, 1986.

[93] Cohn, S. B., "Microwave Bandpass Filters Containing High-Q Dielectric Resonators," *IEEE Trans. Microwave Theory & Tech.,* Vol. MTT-16, April 1968, pp. 218–227.

[94] Harrison, W. H., "A Miniature High-Q Bandpass Filter Employing Dielectric Resonators," *IEEE Trans. Microwave Theory & Tech.,* Vol. MTT-16, April 1968, pp. 210–218.

[95] ESA/ESTEC Workshop, Microwave Filters for Space Applications, June 7–8, 1990, Noordwijk, The Netherlands.

[96] Wakino, K., T. Nishikawa, and Y. Ishikawa, "Miniaturization Techniques of Dielectric Resonator Filters for Mobile Communications," *IEEE Trans. Microwave Theory & Tech.,* Vol. MTT-42, July 1994, pp. 1295–1300.

[97] NTK Production Guide, NTK Piezoelectric Ceramics Div. Eng., Japan, 1997.

[98] Tekelec Production Guide, Tekelec Components, France, 1997.

[99] Flory, A., and R. C. Taber, "High Performance Distributed Bragg Reflector Microwave Resonator," *IEEE Trans. Ultrasonics, Feroelectronics, and Freq. Control,* Vol. 44, March 1997, pp. 486–495.

[100] Alpha/Trans-Tech Production Guide, Alpha/Trans-Tech, Inc, USA, 1997.

[101] Murata Production Guide, Murata Manufacturing Co., Ltd., Japan, 1997.

Selected Bibliography

Gardiol, F. H., "Higher-Order Modes in Dielectrically Loaded Rectangular Waveguides," *IEEE Trans. Microwave Theory & Tech.*, Vol. MTT-16, November 1968, pp. 919–924.

Vartanian, P. H., W. P. Ayres, and A. L. Helgesson, "Propagation in Dielectric Slab Loaded Rectangular Waveguide," *IRE Trans. Microwave Theory & Tech.*, Vol. MTT-6, April 1958, pp. 215–222.

3

Characterization of Discontinuities

This chapter discusses the electromagnetic field analysis of discontinuities in E-plane, ridged waveguide, and CPW filters. An electromagnetic field analysis predicts the electrical behavior of discontinuities with a specified geometry by solving Maxwell's equations for the appropriate boundary conditions. The geometry of the commonly used discontinuity employed in conventional (rectangular waveguide) E-plane filters such as the metal septum in rectangular waveguide is shown in Figure 3.1. Some of the commonly used discontinuities in ridged waveguide E-plane filters such as the metal septum in ridged waveguide, the metal septum between two ridged waveguides with different gaps and the metal septum between rectangular and ridged waveguide are shown in Figure 3.2(a). In this book, only a finite length septum is considered. The septum is assumed to be located at the center of the guide and lossless. Figure 3.2(b) shows the geometry of the CPW coupling gap, a commonly used discontinuity employed in CPW filters.

Electromagnetic analysis of the septum in rectangular waveguide is relatively simple, while the analysis of the septum in ridged waveguide is more

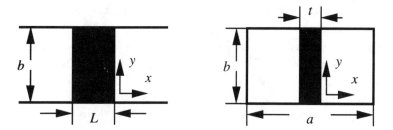

Figure 3.1 Discontinuity step rectangular waveguide to bifurcated rectangular waveguide.

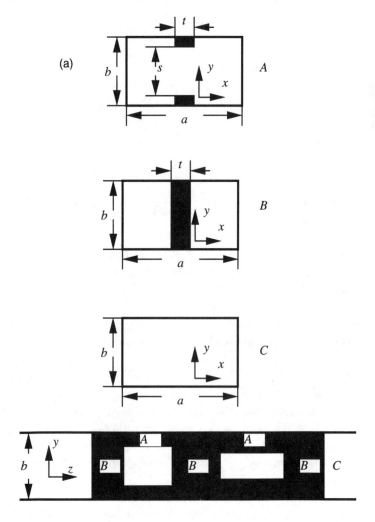

Figure 3.2 (a) Step-ridged waveguide to bifurcated rectangular waveguide discontinuity and (b) geometry of the coplanar waveguide coupling gap.

complicated. The major complexity of this type of discontinuity is in determining the field distribution of the eigenmodes in the ridged waveguide region. To carry out accurate design of E-plane filters, it is important to characterize the discontinuities that are required to be incorporated in them.

E-plane discontinuities have been studied using different rigorous methods, including the variational method [1–3], general purpose methods [4,5], and the mode-matching method [6–9]. A good choice of the numerical methods is the product of tradeoffs among accuracy, speed, storage requirement, versatility, and so on, and is quite structure dependent. General purpose methods,

(b)

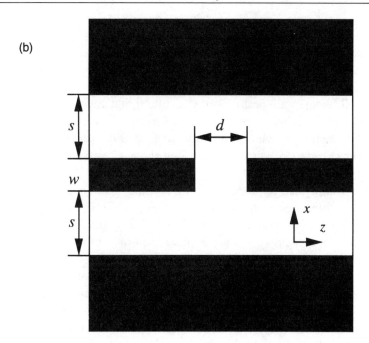

Figure 3.2 (continued).

for instance the finite-element method [4,5], can be applied to problems with nearly arbitrary geometries or structures of complicated shape. Postprocessing must be done to obtain the S parameters. A serious drawback, particularly with the variational method, is that the formulation for a given structure may call for considerable mathematical ability on the part of the user.

However, among these numerical methods, the mode-matching, or modal, analysis originally presented by Wexler [7] proves faster and more efficient for electromagnetic analysis of E-plane discontinuities [10]. The main advantage of this method is its easy implementation on a modern computer system. Mode matching is one of the oldest, most popular, and most frequently employed rigorous full-wave methods to solve the scattering problem due to various discontinuities in waveguide [6,7,11].

The first step in the mode-matching procedure entails the expansion of unknown fields in the individual regions in terms of their respective normal modes. Because the functional form of the normal modes is known, the problem reduces to that of determining the set of modal coefficients (amplitudes) associated with the field expansions in various regions. That procedure, in conjunction with the orthogonality property of the normal modes, leads to a set of linear simultaneous equations for the unknown modal coefficients. The equations then are set up by enforcing the continuity condition for the tangential electric and magnetic fields. The introduction of a matrix representation and the particular choice of the family of modes helps simplify the mathematical efforts and reduce computational time. This method has several different possible formulations, all theoretically equivalent (mathematically valid), although they may be different numerically because of the different manipulations of the basic equations and their individual implementation in a software package. Because of the singular behavior of the magnetic field at the edges of the septa, a large number of modes need to be included in the field expansions to ensure good convergence.

The application of the mode-matching method to the electromagnetic analysis of discontinuities in conventional E-plane filters (metal septum in rectangular waveguide) is discussed in Section 3.1. Simplifications that arise when evanescent modes between adjacent discontinuities do not interact and the convergence problem also are covered. Section 3.2 analyzes discontinuities in ridged waveguide E-plane filters, such as the metal septum between two ridged waveguides with identical and different gaps, and metal septa between rectangular and ridged waveguides. That same section also examines ridged waveguide, the convergence problem, and results of the analysis of the septum in ridged waveguide associated with the numerical mode-matching method. Section 3.3 introduces electromagnetic analysis of the CPW discontinuities such as CPW coupling gap [see Figure 3.2(b)] and edge-coupled CPW lines by Em^{TM} software [12].

3.1 Analysis of Filter Discontinuities by the Mode-Matching Method

3.1.1 Discontinuities in Conventional E-Plane Filters

This section applies the mode-matching method to the analysis of discontinuities in conventional E-plane filters such as the metal septum in rectangular waveguide. The septum is assumed to be of finite thickness, of finite length, and lossless. The geometry of the problem and the coordinate system used are shown in Figure 3.3. The septum is assumed to be centered in the waveguide with the energy propagating in the z direction. The approach used in this section is based on the modal analysis method of Wexler [7].

3.1.1.1 Metal Septum in Rectangular Waveguide

To apply the modal analysis method, we consider the standard waveguide, denoted as region a in Figure 3.3, while the reduced waveguide is denoted as region b. In view of the longitudinal symmetry, we can simplify the problem,

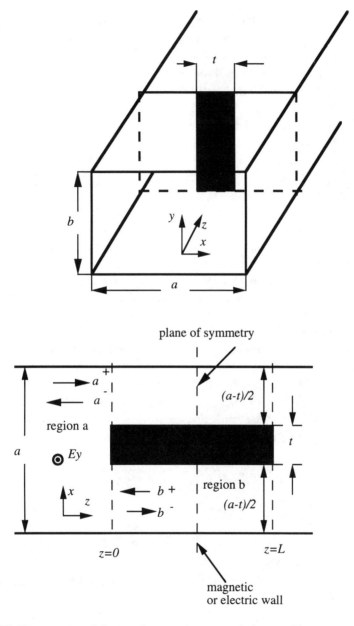

Figure 3.3 The geometry of the metal septum in rectangular waveguide.

considering even/odd excitation corresponding to placing a magnetic/electric wall at $z = L/2$. Therefore, a symmetrical septum can be analyzed by considering the left half of the structure, with the symmetry plane representing an open circuit for even excitation and a short circuit for odd excitation. When applying the mode-matching method, it is important to know which components of the electromagnetic field are involved. They are determined by the field components of the exciting mode (e.g., the dominant waveguide mode) plus those added by all discontinuities. Because usually all components tangential to the discontinuity plane need to be matched, the number of field components directly affects the matrix sizes and the CPU time required for the analysis of a discontinuity.

At the discontinuity plane, a dominant mode TE_{10} excitation will not introduce field components other than those of the incident wave such as E_y, H_x, H_z, because the discontinuity is uniform along the y-axis. That discontinuity can be characterized by TE_{m0} modes, where m has all possible positive odd integers, 1, 3, 5, . . . in standard waveguide, while m can take on all positive integers, $m = 1, 2, 3, . . .$, in the reduced waveguide region. Due to the singular behavior of the magnetic field at the edge of the metal septum in rectangular waveguide, numerical problems can arise.

Consider a single propagating mode with mode coefficient a_1 emanating from a matched source in guide a, and impinging on reduced waveguide b at junction $z = 0$. At junction $z = 0$, let ρ be the reflection coefficient of this mode, and let a_i, where $i = 2, 3, . . .$ be the mode coefficients of the scattered modes. The total transverse fields that are just to the left of junction $z = 0$ (in region a) can be expressed in terms of those modes to yield

$$E_{ta} = (1 + \rho)a_1 e_{a1} + \sum_{i=2}^{M} a_i e_{ai} \tag{3.1a}$$

$$H_{ta} = (1 - \rho)a_1 h_{a1} - \sum_{i=2}^{M} a_i h_{ai} \tag{3.1b}$$

Just to the right of junction $z = 0$ (in region b), the total transverse fields can be expressed as

$$E_{tb} = \sum_{j=1}^{N} b_j \left(e_{bj} + \sum_{k=1}^{N} s_{jk} e_{bk} \right) \tag{3.2a}$$

$$H_{tb} = \sum_{j=1}^{N} b_j \left(h_{bj} - \sum_{k=1}^{N} s_{jk} h_{bk} \right) \tag{3.2b}$$

Expressions for the transverse fields in region a are

$$e_{ai} = \sin\left(\frac{p\pi x}{a}\right) \tag{3.3a}$$

$$h_{ai} = Y_{ai} \sin\left(\frac{p\pi x}{a}\right) \tag{3.3b}$$

The wave admittance of the ith mode is defined by

$$Y_{ai} = \sqrt{\frac{\epsilon_0}{\mu_0}} \sqrt{1 - \left(\frac{p\lambda_0}{2a}\right)^2} \tag{3.4}$$

where λ_0 is the wavelength in free space at f_0, and $p = 2i - 1$. Modes are numbered consecutively, that is, $i = 1, 2, 3, \ldots, M$.

Expressions for the transverse fields in region b are

$$e_{bj} = \sin\left(\frac{2q\pi x}{a - t}\right) \tag{3.5a}$$

$$h_{bj} = Y_{bj} \sin\left(\frac{2q\pi x}{a - t}\right) \tag{3.5b}$$

Equation (3.2) holds in the range $0 < x < (a - t)/2$ and is zero across the vane. The wave admittance of the jth mode is

$$Y_{bj} = \sqrt{\frac{\epsilon_0}{\mu_0}} \sqrt{1 - \left(\frac{q\lambda_0}{a - t}\right)^2} \tag{3.6}$$

where $q = 1, 2, \ldots, N$. One should substitute $(1 - x/a)$ for (x/a) when $(a + t)/2 < x < a$.

To solve for the unknown parameters p, a_i, and b_j, the boundary conditions at the discontinuity must be satisfied. Boundary conditions to be satisfied at the discontinuity are as follows: Continuity of the total transverse electric and magnetic fields across the aperture and the zero tangential electric field at the surface that contains the septum. We presume that s_{jk} due to the second discontinuity is known; otherwise, it must be evaluated by solving the second discontinuity first. That leads to tedious computations, which can be avoided for this problem because the discontinuity is symmetrical about the $z = L/2$ plane. By using symmetrical and asymmetrical excitations, we can find the normalized reactive impedances (z_e, z_0) without a complete set of scattering coefficients. Symmetrical excitation of the septum is obtained by having two modes in the waveguide, one traveling in the $+z$ direction and the other in the $-z$ direction, such that the E fields are in phase in the $z = L/2$ plane; antisymmetrical excitation is obtained if the modes are 180 degrees out of phase. For symmetrical excitation, an open circuit appears at the symmetry plane; antisymmetrical excitation produces a short circuit. With the open circuit at the central plane (Figure 3.4), with $s_{jk} = +1$, the normalized reactive even-mode impedance z_e is given by

$$z_e = \frac{1 + \rho_e}{1 - \rho_e} \qquad (3.7)$$

With the short circuit at the central plane (see Figure 3.4), with $s_{jk} = -1$, the normalized reactive odd-mode impedance z_0 is given by

$$z_0 = \frac{1 + \rho_0}{1 - \rho_0} \qquad (3.8)$$

Mathematical details of the derivations' normalized reactive impedances (z_e, z_0) are given in Appendix 3A.

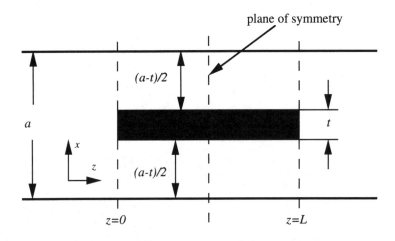

Figure 3.4 Symmetrical finite length metal septum in rectangular waveguide.

3.1.1.2 Simplification When Evanescent Modes Between Adjacent Discontinuities Do Not Interact

In conventional E-plane filters, if the length, l, of the resonator is comparable to the internal waveguide width, a, the effect of higher-order mode coupling along the filter section (resonator) can be neglected. Under those conditions, only the dominant mode is responsible for the coupling between septa (discontinuities). In the case of longitudinally symmetric structures, such as the conventional E-plane bandpass filters under consideration, we can significantly simplify the analysis. In that case, the structure is analyzed by placing electric and magnetic walls at the plane of symmetry. By using Bartlett's bisection theorem [13], the insertion loss (L_I) and return loss (L_R) of a symmetrical E-plane filter can be expressed as

$$L_I = 20 \log_{10} \frac{(1 + z_o)(1 + z_e)}{z_e - z_o} \tag{3.9}$$

and

$$L_R = 20 \log_{10} \frac{(1 + z_o)(1 + z_e)}{1 - z_e z_o} \tag{3.10}$$

where $j z_{e(o)}$ is the normalized input impedance of the two identical 1-ports formed by placing a magnetic (electric) wall at the plane of symmetry. $z_{e(o)}$ can be calculated by transforming an open (short) circuit placed at the plane of symmetry through the filter sections (resonators and E-plane septa) located to the left of the plane of symmetry. Each E-plane septum is itself symmetrical and can be electrically represented by normalized even- and odd-mode impedances

$$z_{ei} = j(x_{si} + 2x_{pi}) \tag{3.11}$$

$$z_{oi} = j x_{si} \tag{3.12}$$

where x_{si} and x_{pi} are the normalized reactances of a symmetrical normalized equivalent T circuit of the metal septum in waveguide (Figure 3.5).

For a normalized reactive impedance jz, an E-plane septum performs the normalized impedance transformation $jz \Rightarrow j z_{in}$, where z_{in} is given by

$$z_{in} = \frac{z(z_{ei} + z_{oi}) + 2 z_{ei} z_{oi}}{2z + (z_{ei} + z_{oi})} \tag{3.13}$$

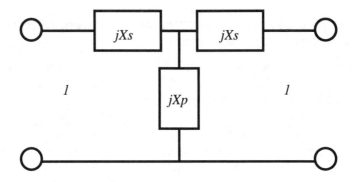

Figure 3.5 A symmetrical normalized equivalent T circuit of the metal septum.

A resonator section, that is, a length of guide, performs the normalized impedance transformation $jz \Rightarrow jz_{in}$, where z_{in} is given by

$$z_{in} = \frac{z + \tan \beta l}{1 - z \tan \beta l} \tag{3.14}$$

where $\beta (= 2\pi/\lambda_g)$ is the propagation constant, and l is the length of the resonator. By applying (3.13) and (3.14), it is possible to calculate z_e and z_o starting at the center of the filter and working outward. The process involves no matrix manipulation and uses only real arithmetic.

3.1.2 Discontinuities in Ridged Waveguide Filters

This section applies the mode-matching method to the analysis of the discontinuities in ridged waveguide E-plane filters. At the discontinuity plane of the conventional E-plane filters discussed so far, a dominant mode TE_{10} excitation does not introduce field components other than those of the incident wave, such as E_y, H_x, H_z. In contrast, in the case of discontinuities in E-plane ridged waveguide filters, at the discontinuity plane shown in Figure 3.2, a dominant mode TE_{10} excitation introduces field components other than those of the incident wave, such as E_x, E_y, H_x, H_y, H_z. These types of discontinuities can be characterized by TE_{mn} modes only where m has all odd possible positive values, 1, 3, 5, . . . , and n all even possible positive values, 0, 2, 4, 6, . . . in the ridged waveguide region (region a), while m and n have all possible positive values, 1, 2, 3, . . . in the reduced waveguide region (region b) and need to be considered.

First, a number of TE modes will be evaluated in the ridged waveguide sections as a function of the geometrical parameters of the guide. This is performed by applying the mode-matching method along the transversal direction. The cutoff frequencies of ridged waveguide modes are found by an iterative algorithm that searches for the zeros of a function corresponding to transverse resonance conditions [14]. Then the scattering parameters of several discontinuities in ridged waveguide filters are determined.

3.1.2.1 Analysis of Ridged Waveguide

Figure 3.6 illustrates the cross-sectional shape of the ridged waveguide with the coordinates, where t denotes the thickness of the ridge, s the gap in the ridge, 1 the region of $0 < x < t/2$, and 2 the region of $t/2 < x < a/2$. The width and the height of the waveguide cross-section are given by a and b, respectively. Considering the odd TE modes, whose symmetry plane has been assumed to be the y coordinate axis ($x = 0$) in this analysis, the $x = 0$ plane can be regarded as a magnetic wall. In the case of even modes, a similar analysis can be accomplished, regarding the $x = 0$ plane as an electric wall. Therefore, attention will be focused on TE modes (fields) with a magnetic symmetry plane about the y coordinate axis ($x = 0$), and only the fields in regions 1 and 2 need to be characterized.

The TE fields in ridged waveguide can be derived from the magnetic Hertzian potential Q, which satisfies the Helmholtz equation in the transverse plane:

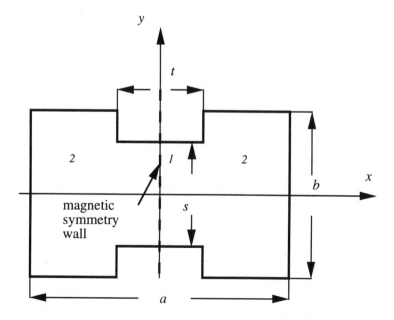

Figure 3.6 Geometry of ridged waveguide.

$$\Delta_t^2 Q + k_c^2 Q = 0 \tag{3.15}$$

with

$$k_c^2 = \omega^2 \mu_0 \epsilon_0 - \beta^2 \tag{3.16}$$

where β is the propagation constant, ω is the angular frequency, μ_0 is the permeability of free space, and ϵ_0 is the permittivity of free space. The scalar potential function is related to the electric and magnetic fields [15] by the equations

$$E_t = e_z \cdot \Delta_t Q \tag{3.17}$$

$$H_t = - \frac{\beta}{\omega \mu_0} \Delta_t Q \tag{3.18}$$

where e_z is a unit vector of the z direction. Applying the TE boundary conditions, we find the following expressions for Q.

In region 1 $(0 < x < t/2)$,

$$Q_1 = \sum_{n=0}^{N} \eta_{1n} \sin(k_{x1n}x) \cos\left[k_{y1n}\left(y - \frac{s}{2} \right) \right] \tag{3.19}$$

where

$$k_{x1n} = \sqrt{k_c^2 - k_{y1n}^2} \qquad k_{y1n} = \frac{n\pi}{s} \tag{3.20}$$

In region 2 $(t/2 < x < a/2)$,

$$Q_2 = \sum_{m=0}^{M} \eta_{2m} \sin\left[k_{x2m}\left(x - \frac{a}{2} \right) \right] \cos\left[k_{y2m}\left(y - \frac{b}{2} \right) \right] \tag{3.21}$$

where

$$k_{x2m} = \sqrt{k_c^2 - k_{y2m}^2} \qquad k_{y2m} = \frac{m\pi}{b} \tag{3.22}$$

Performing the operations in (3.17) and (3.18) on (3.19) and (3.21), we find

$$E_{t1} = -\sum_{n=0}^{N} \eta_{1n} \left\{ k_{y1n} \sin(k_{x1n}x) \sin\left[k_{y1n}\left(y - \frac{s}{2}\right)\right] e_x \right.$$

$$\left. + k_{x1n} \cos(k_{x1n}x) \cos\left[k_{y1n}\left(y - \frac{s}{2}\right)\right] e_y \right\} \tag{3.23}$$

in region 1 and

$$E_{t2} = -\sum_{m=0}^{M} \eta_{2m} \left\{ k_{y2m} \cos\left[k_{x2m}\left(x - \frac{a}{2}\right)\right] \sin\left[k_{y2m}\left(y - \frac{b}{2}\right)\right] e_x \right.$$

$$\left. - k_{x2m} \sin\left[k_{kx2m}\left(x - \frac{a}{2}\right)\right] \cos\left[k_{y2m}\left(y - \frac{b}{2}\right)\right] e_y \right\} \tag{3.24}$$

in region 2.

Let $E_{gap}(y) = E_y(t/2, y)$ be the y component of the electric field at the boundary between regions 1 and 2. Because $E_{gap}(y)$ must be continuous across this interface,

$$E_{gap}(y) = -\sum_{n=0}^{N} \eta_{1n}k_{x1n} \cos\left(\frac{k_{x1n}t}{2}\right) \cos\left[\frac{n\pi}{s}\left(y - \frac{s}{2}\right)\right]$$

$$= \sum_{m=0}^{M} \eta_{2m}k_{x2m} \sin\left[k_{x2m}\left(\frac{t-a}{2}\right)\right] \cos\left[\frac{m\pi}{b}\left(y - \frac{b}{2}\right)\right] \tag{3.25}$$

Equation (3.26) yields the Fourier coefficients η_{1n} and η_{2m}, that is,

$$\int_{-s/2}^{s/2} E_{gap}(y) \cos\left[\frac{n\pi}{s}\left(y - \frac{s}{2}\right)\right] dy = -\epsilon_n\eta_{1n}sk_{x1n} \cos\left(\frac{k_{x1n}t}{2}\right) \tag{3.26}$$

$$\int_{-s/2}^{s/2} E_{gap}(y) \cos\left[\frac{m\pi}{b}\left(y - \frac{b}{2}\right)\right] dy = \epsilon_m\eta_{2m}bk_{x2m} \sin\left[k_{x2m}\left(\frac{t-a}{2}\right)\right]$$

$$\tag{3.27}$$

in which ϵ_s is the Neumann number; ($\epsilon_s = 1$ if $s = 0$; otherwise, $\epsilon_s = 2$). Note that in (3.27) the integrations embrace only the interval $-s/2 < y < s/2$ because $E_{gap}(y) = 0$ for $s/2 < y < b/2$. Continuity of the scalar potential in conjunction with (3.26) and (3.27) results in the integral equation

$$\sum_{m=0}^{M} \frac{\cot\left[k_{x2m}\left(\frac{a-t}{2}\right)\right]}{k_{x2m}\epsilon_{m}b} \cos\left[k_{y2m}\left(y-\frac{b}{2}\right)\right] \cdot \int_{-s/2}^{s/2} E_{gap}(y) \cos\left[k_{y2m}\left(y-\frac{b}{2}\right)\right] dy$$

$$= \sum_{n=0}^{N} \frac{\tan\left[k_{x1n}\left(\frac{t}{2}\right)\right]}{k_{x1n}\epsilon_{n}s} \cos\left[k_{x1n}\left(y-\frac{s}{2}\right)\right] \cdot \int_{-s/2}^{s/2} E_{gap}(y) \cos\left[k_{y1n}\left(y-\frac{s}{2}\right)\right] dy$$

$$\tag{3.28}$$

The eigenvalues k_c for the TE mode can be extracted from (3.28). However, the procedure is complicated by the fact that concomitantly one must find $E_{gap}(y)$, the y component of the TE electric field at the region 1/region 2 boundary. An exact analytical solution is not known. However, a numerical solution may be obtained. One technique for doing that is the Ritz-Galerkin method [14]. The Ritz-Galerkin method consists of expanding the unknown boundary field $E_{gap}(y)$ by a suitably chosen family of functions; it then requires that the resultant equation be orthogonal to each expansion function. A matrix equation is obtained and may be solved by suitably developed matrix theory. In the case of ridged waveguide, because the edge of the ridge has a right angle, the y component of an electric field near the edge is approximately proportional to $\Delta y^{-1/3}$ [16], where Δy represents the distance between the edge and an observational point. Therefore, by taking into account the edge conditions mentioned by Mittra [11], it is convenient to expand $E_{gap}(y)$ in the eigenfunctions of region 1:

$$E_{gap}(y) = \sum_{l=0}^{L} C_l \cos\left(\frac{l\pi y}{s}\right)\left[\left(\frac{s}{2}\right)^2 - y^2\right]^{-1/3} \tag{3.29}$$

Substitution of (3.29) into (3.28) and after integration, multiplying through by $\cos\left(\left(\epsilon\frac{\pi}{s}\right)((s/2)-y)\right)$, followed by integration over $[-s/2, s/2]$, yields the matrix equation

$$[H(k_c)] \cdot [C] = [0] \tag{3.30}$$

with the matrix elements H_{ij} given by

$$H_{ij}(k_c) = \delta_{ij}\epsilon_i \cdot s\frac{\tan\left(k_{x1i}\frac{t}{2}\right)}{k_{x1i}} - \sum_{m=0}^{M} P_{im}P_{jm}\frac{\cot\left[k_{x2m}\left(\frac{a-t}{2}\right)\right]}{k_{x2m}\epsilon_m b} \qquad (3.31)$$

where

$$P_{im} = \int_{-s/2}^{s/2} \cos\frac{i\pi}{s}\left[\left(\frac{s}{2}\right)^2 - y^2\right]^{-1/3}\cos\left[\frac{m\pi}{b}\left(y - \frac{b}{2}\right)\right]dy \qquad (3.32)$$

$$P_{jm} = \int_{-s/2}^{s/2} \cos\frac{j\pi}{s}\left[\left(\frac{s}{2}\right)^2 - y^2\right]^{-1/3}\cos\left[\frac{m\pi}{b}\left(y - \frac{b}{2}\right)\right]dy \qquad (3.33)$$

and

$$[C] = [C_0 C_1 C_2 \ldots C_L]^t \qquad (3.34)$$

In (3.31), Δ_{ij} is the Kronecker delta ($\Delta_{ij} = 1$ if $i = j$; $\Delta_{ij} = 0$ if i_j). Equation (3.30) is a generalized matrix eigenvalue problem. The TE eigenvalues k_c are solutions of the nonlinear equation

$$\det|H(k_c)| = 0 \qquad (3.35)$$

The convergence of the dominant as well as higher order eigenvalues is discussed in subsection 3.2.1.1. The vector $[C]$, which is a solution of (3.30) for a particular k_c, is the associated eigenvector.

3.1.2.2 Metal Septum in Rectangular Waveguide Between Two Different Ridged Waveguides

This section analyzes the metal septum in rectangular waveguide between two different ridged waveguide by the mode-matching method. The original mode-matching method formulation [5] is adapted to the case of the metal septum in ridged waveguide. The metal septum is assumed to be located at the center of the guide to reduce the excitation of the even-order modes, and the structure is assumed to be lossless, as shown in Figure 3.7.

Basically, the analysis consists of two parts. The first part deals with a scattering problem in a ridged waveguide bifurcated by a septum of finite thickness. In the second part, two junctions, as in the first part, are joined back-to-back to form a finite-width septum. The process takes into account the interaction between junctions of both the dominant mode and higher order modes.

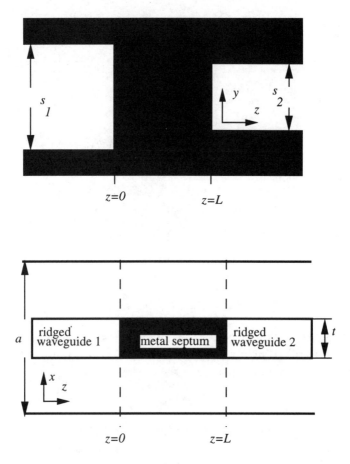

Figure 3.7 Metal septum inserted in rectangular waveguide between two different ridged waveguides.

Ridged Waveguide: Metal Septum Step

Consider the step-ridged waveguide to bifurcated rectangular waveguide discontinuity shown in Figure 3.8, where the septum is located at the center along the E-plane. Because of symmetry, we can place a magnetic wall at the center along the z-axis and consider only one-half of the structure. The problem becomes a transverse step junction between ridged waveguide and rectangular waveguide with widths $a/2$ and $(a - t)/2$, respectively, as shown in Figure 3.8(b). Assuming excitation by the TE_{10} dominant mode, the only types of higher order modes that are excited (generated) at the step discontinuity are TE_{mn} modes, where m has all odd possible positive values, 1, 3, 5, . . . , $n = 0, 2, 4, . . .$ in the ridged waveguide region, and $m = 1, 2, 3, . . .$ and $n = 0, 1, 2, 3, . . .$ in the reduced waveguide region. Due to the singular

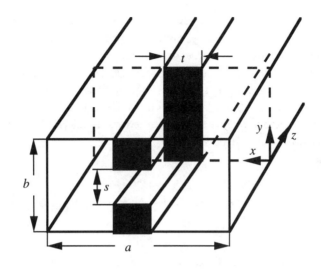

(a)

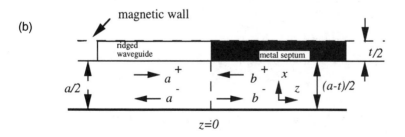

(b)

Figure 3.8 Configuration for the rigorous field theory treatment: (a) forward and backward waves at the discontinuity step and (b) ridged waveguide to bifurcated rectangular waveguide step with magnetic wall symmetry.

behavior of the fields at the edges of the bifurcated rectangular waveguide and the ridges in ridged waveguide, numerical problems can arise.

The mode-matching procedure begins with expanding the tangential components of the electric and magnetic fields at the discontinuity in terms of the normal modes in each region. The amplitudes of the normal modes are chosen such that the boundary conditions are satisfied at the discontinuity.

With reference to Figure 3.7(b), the total transverse fields in each region at the discontinuity can be written [6] as:

region $a(z < 0)$

$$E_{ta} = \sum_{m=1}^{M} \phi_{am}[a_m^+ \exp(-j\beta_{am}z) + a_m^- \exp(+j\beta_{am}z)] \qquad (3.36)$$

$$H_{ta} = \sum_{m=1}^{M} Y_{am}\phi_{am}[a_m^+ \exp(-j\beta_{am}z) - a_m^- \exp(+j\beta_{am}z)] \qquad (3.37)$$

where

$$\beta_{am}^2 = \omega^2\mu_0\epsilon_0 - k_c^2 \qquad (3.38)$$

and

$$Y_{am} = \frac{\beta_{am}}{\omega\mu_0} \qquad (3.39)$$

The normalized mode functions operating in (3.36) and (3.37) can be obtained from the eigenmode analysis of the ridged waveguide. In the case of the reduced waveguide, in region b ($z > 0$, $0 < x(a - t)/2$), the fields are given by

$$E_{tb} = \sum_{p=1}^{K} \phi_{bp}\{[b_p^+ \exp(-j\beta_{bp}z) + b_p^- \exp(+j\beta_{bp}z)]\} \qquad (3.40)$$

$$H_{tb} = \sum_{p=1}^{K} Y_{bp}\phi_{bp}\{[b_p^+ \exp(-j\beta_{bp}z) + b_p^- \exp(+j\beta_{bp}z)]\} \qquad (3.41)$$

where

$$\beta_{bp}^2 = \omega^2\mu_0\epsilon_0 - k_{cbp}^2 \qquad (3.42)$$

$$k_{cbp}^2 = \left(\frac{2m\pi}{a-t}\right)^2 + \left(\frac{n\pi}{b}\right)^2 \tag{3.43}$$

$$Y_{bp} = \frac{\beta_{bp}}{\omega\mu_0} \tag{3.44}$$

and

$$\phi_{bp} = A_p \frac{\omega\mu_0}{k_{cbp}^2}\left[\left(\frac{n\pi}{b}\right)\cos\left(\frac{2m\pi x}{a-t}\right)\sin\left(\frac{n\pi y}{b}\right)\right]$$

$$- A_p \frac{\omega\mu_0}{k_{cbp}^2}\left[\left(\frac{2m\pi}{a-t}\right)\sin\left(\frac{2m\pi x}{a-t}\right)\cos\left(\frac{n\pi y}{b}\right)\right] \tag{3.45}$$

with

$$A_p = \frac{k_{cbp}\sqrt{2\epsilon_m\epsilon_n}}{\sqrt{(a-t)b\omega\mu_0\beta_p}} \qquad \epsilon_0 = 1 \; \epsilon_m(m>1)=2 \tag{3.46}$$

where the indices p are related to the waveguide modes m,n by rearranging them with respect to increasing cutoff frequencies; k_c is a ridged waveguide eigenvalue that is related to the cutoff frequency, which can be easily calculated by solving the related eigenvalue problem; ϕ_{am} and ϕ_{bp} are the normalized mode functions in regions a and b, with phase propagation constants b_{am} and b_{bp}, respectively; Y_{am} and Y_{bp} are wave admittances in regions a and b; and a_m^+ and b_p^- are the amplitudes of the waves traveling in the positive (forward) z-direction for regions a and b, while a_m^- and b_p^+ are the amplitudes of the waves traveling in the negative (backward) z-direction in regions a and b, respectively. It should be noted that according to the voltages and currents on transmission lines, the electric and magnetic field components are given as the sum and the difference of the forward and backward traveling waves, respectively. The boundary conditions necessary for matching the field components at the discontinuity ($z = 0$) are:

$$E_{ta} = 0 \qquad \frac{a-t}{2} \le x \le \frac{a+t}{2}$$

$$E_{ta} = E_{tb} \qquad 0 \le x \le \frac{a-t}{2}$$

$$H_{ta} = H_{tb} \qquad 0 \le x \le \frac{a-t}{2} \tag{3.47}$$

and from (3.36), (3.37), (3.40), and (3.41), it follows that

$$\sum_{m=1}^{M} (a_m^+ + a_m^-)\phi_{am} = \sum_{p=1}^{K} (b_p^+ + b_p^-)\phi_{bp} \quad 0 \le x \le \frac{a-t}{2} \quad (3.48)$$

$$\sum_{m=1}^{M} (a_m^+ - a_m^-)Y_{am}\phi_{am} = \sum_{p=1}^{K} (b_p^+ - b_p^-)Y_{bp}\phi_{bp} \quad 0 \le x \le \frac{a-t}{2} \quad (3.49)$$

By using the property of mode orthogonality, we can derive a system of equations involving the unknown coefficients from (3.48) and (3.49). The next step is to eliminate the x dependence in (3.48) and (3.49). Following Wexler's procedure, we enforce continuity of the transverse electromagnetic field at junction ($z = 0$) by (3.48) and (3.49), taking the cross-product with ϕ_{am}, and integrating with respect to x and y from 0 to $(a - t)/2$ and 0 to b, respectively, we find

$$a_m^+ + a_m^- = \sum_{p=1}^{K} H_{mp}(b_p^+ + b_p^-) \quad m = 1, 2, \ldots, M \quad (3.50)$$

$$Y_{am}(a_m^+ + a_m^-) = \sum_{p=1}^{K} Y_{bm}H_{mp}(b_p^+ + b_p^-) \quad m = 1, 2, \ldots, M \quad (3.51)$$

This is a first set of linear equations, where H_{mn} is a coupling integral defined at the plane of the discontinuity as

$$H_{mp} = \int_{0}^{(a-t)/2} \int_{0}^{b} \phi_{am}\phi_{bp} \, dx \, dy \quad (3.52)$$

A second set of linear equations can be obtained by multiplying (3.5a) and (3.5b) by ϕ_{bp} and integrating with respect to x and y from 0 to $(a - t)/2$ and 0 to b, respectively. This yields

$$\sum_{p=1}^{M} H_{pm}(a_p^+ + a_p^-) = b_m^+ + b_m^- \quad m = 1, 2, \ldots, K \quad (3.53)$$

$$\sum_{p=1}^{M} H_{pm}Y_{ap}(a_p^+ - a_p^-) = Y_{bm}(b_m^+ - b_m^-) \quad m = 1, 2, \ldots, K \quad (3.54)$$

From this point, several approaches are possible. It is easier to handle (3.50) to (3.54) simultaneously in matrix form. The matrix form of those equations is written as:

$$E_t: a^+ + a^- = H(b^+ + b^-) \tag{3.55}$$

$$H_t: a^+ - a^- = Y_a^{-1} H Y_b (b^+ - b^-) \tag{3.56}$$

$$E_t: H^t(a^+ + a^-) = b^+ + b^- \tag{3.57}$$

$$H_t: Y_b^{-1} H^t(a^+ - a^-) = b^+ - b^- \tag{3.58}$$

where H is a matrix of size M by K with element H_{mp} as defined by (3.52). The superscript t denotes the transpose operation, and $Y_i (i = a,b)$ are diagonal matrices with diagonal elements Y_{am} and Y_{bp}. The coefficient vectors are defined as

$$a^+ = \begin{bmatrix} a_1^+ \\ a_2^+ \\ \vdots \\ a_M^+ \end{bmatrix} \qquad b^- = \begin{bmatrix} b_1^- \\ b_2^- \\ \vdots \\ b_K^- \end{bmatrix} \tag{3.59}$$

$$a^- = \begin{bmatrix} a_1^- \\ a_2^- \\ \vdots \\ a_M^- \end{bmatrix} \qquad b^+ = \begin{bmatrix} b_1^+ \\ b_2^+ \\ \vdots \\ b_K^+ \end{bmatrix} \tag{3.60}$$

a^+ and b^- are column vectors of the excitation terms, and a^- and b^+ are column vectors of unknown modal coefficients. Two independent vectors are required to solve for two unknown vectors. Hence, for four pairs of equations [i.e., (3.55) and (3.56), (3.56) and (3.57), (3.57) and (3.58), and (3.58) and (3.55)], substituting one equation into the other in the same pair, we have eight ways of obtaining a solution for a^+ and b^-.

The final solutions are expressed by scattering parameters defined by

$$\begin{bmatrix} a^- \\ b^+ \end{bmatrix} = \begin{bmatrix} S_{11} & S_{12} \\ S_{21} & S_{22} \end{bmatrix} \begin{bmatrix} a^+ \\ b^- \end{bmatrix} \tag{3.61}$$

where S_{ij} ($i = 1,2$; $j = 1,2$) are the generalized scattering parameters (submatrices) representing the amplitude of the scattered field at port i due to the unit incident field at port j. The detailed expressions for scattering parameters are included in Appendix 3B. In this research work, the formulation [see (3.62) to (3.65)] was employed because of the best satisfaction of the boundary conditions [7], which is given by

$$S_{22} = [I + Y_b^{-1} H^t Y_a H]^{-1} [I - Y_b^{-1} H^t Y_a H] \tag{3.62}$$

$$S_{21} = 2[I + Y_b^{-1} H^t Y_a H]^{-1} Y_b^{-1} H^t Y_a \tag{3.63}$$

$$S_{12} = H[I + S_{22}] \tag{3.64}$$

$$S_{11} = H S_{21} - I \tag{3.65}$$

Finite-Length Septum

With the knowledge of the scattering parameters for a single discontinuity (junction), the overall composite scattering matrix can be obtained by a network constitution in terms of the generalized scattering matrices [6]. Let S^1 and S^2 represent the scattering matrices for the isolated junctions at $z = 0$ and $z = L$, respectively (Figure 3.9).

The combination of matrices S^1, T, and S^2 results in the composite scattering matrix S for the overall structure given by (3.66a) to (3.66d) below for the septum of length L, with reference planes located at $z = 0$ and $z = L$.

$$S_{11} = S_{11}^1 + S_{12}^1 T S_{11}^2 (I - T S_{22}^1 T S_{11}^2)^{-1} T S_{21}^1 \tag{3.66a}$$

$$S_{12} = S_{12}^1 (I - T S_{22}^1 T S_{11}^2)^{-1} T S_{12}^2 \tag{3.66b}$$

$$S_{21} = S_{21}^2 (I - T S_{11}^2 T S_{22}^1)^{-1} T S_{21}^1 \tag{3.66c}$$

$$S_{22} = S_{22}^2 + S_{21}^2 T S_{22}^1 (I - T S_{11}^2 T S_{22}^1)^{-1} T S_{12}^2 \tag{3.66d}$$

In (3.66a) to (3.66d), I is the unity matrix, and the transmission matrix T represents the wave propagating (for propagating modes) or attenuating (for evanescent modes) for a distance of L in guide region b. This is a diagonal matrix whose diagonal elements are

$$T(n,n) = e^{(-j\beta_n L)} \qquad n = 1, 2, 3, \ldots \tag{3.67}$$

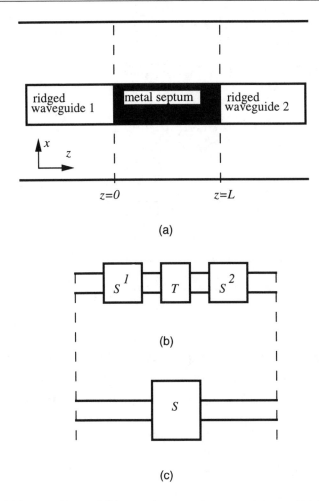

Figure 3.9 (a) Septum of length *L*; (b) scattering network representation; and (c) the composite scattering matrix after network combination.

So that the mode-matching formulation can be employed, it must be possible to split the structure to be analyzed into well-defined regions for which the mode functions are known. For a given discontinuity, all the mode types that can be excited by it must be included so the field continuity equations can be satisfied. The program structure is shown in Figure 3.10. It can be seen that once written, the program can be easily modified for the analysis of different structures. All that is necessary is that a subroutine be written for computation of the required mode functions that returns the normalization factors and propagation constants for the required numbers and types of modes. The array of coupling integrals *H* also must be computed. The remaining parts of the program are unchanged.

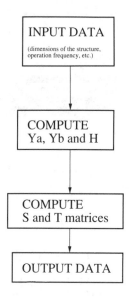

Figure 3.10 The program structure.

3.1.2.3 Metal Septum in Rectangular Waveguide Between Rectangular and Ridged Waveguides

Using the procedure of field matching at $z = 0$ and $z = L$, the elements of the S matrix of the structure shown in Figure 3.11 are given by (3.62) to (3.65). Because the characteristics of those two discontinuities are essentially the same except for the different waveguide eigenvalues (k_c's), we obtain the scattering parameters for both matrices from (3.62) to (3.65). From (3.66a) to (3.66d), we obtain the element values for the overall structure.

3.1.2.4 Metal Septum in Ridged Waveguide

Using the procedure of field matching at $z = 0$ and $z = L$, the elements of the S matrix of the structure shown in Figure 3.12 are given by (3.62) to (3.65). Because the characteristics of those two discontinuities are essentially the same except for the different ridged waveguide eigenvalues (k_c's), we obtain the scattering parameters for both matrices from (3.62) to (3.65). From (3.66a) to (3.66d), we obtain the element values for the overall structure.

3.2 Convergence

For the electromagnetic analysis of E-plane septa in a rectangular waveguide, the mode-matching method is a popular choice. However, due to the singular

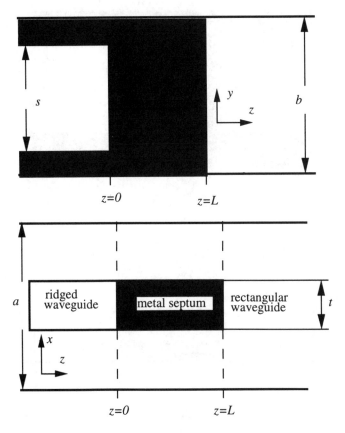

Figure 3.11 Metal septum between rectangular and ridged waveguides.

behavior of the magnetic field at the edges of the septa in a rectangular waveguide, a large number of modes need to be included in the field expansions to ensure good convergence. That point is illustrated by Tables 3.1 and 3.2. Table 3.1 shows the slow convergence of the normalized even- and odd-mode impedances of an E-plane septum with a length of $0.8146a/p$ and a thickness of $0.01094a$ at a center frequency of $1.4478f_c$ (f_c = TE_{10} cutoff frequency in rectangular waveguide) as the number of the modes used in the field expansions is increased. Table 3.2 shows the convergence of the insertion loss of a symmetrical X-band fifth-degree E-plane bandpass filter with an insert thickness of $0.01094a$, septa lengths given by $d_1 = 0.1528a/\pi$, $d_2 = 0.8146a/\pi$, $d_3 = 0.9985a/\pi$, and resonator lengths given by $l_1 = 2.2055a/\pi$, $l_2 = 2.2568a/\pi$, and $l_3 = 2.2603a/\pi$, at a center frequency of $1.4453f_c$, where f_c is the TE_{10} cutoff frequency in rectangular waveguide. The calculation of this insertion loss requires the evaluation of the normalized even- and odd-mode impedances of

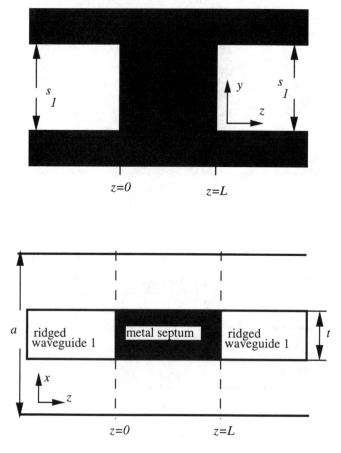

Figure 3.12 Metal septum inserted in rectangular waveguide between two equal ridged waveguides.

three E-plane septa. In all three cases, the same number of modes is used. The calculations were performed on a SUN workstation (SPARC 10). About 0.5 minute of CPU time is required for the calculation of the insertion loss at one frequency using 100 modes.

Tables 3.3 and 3.4 show the slow convergence of the dominant and the higher order eigenvalues of a double-ridged waveguide with b/a = 0.80, s/b = 0.275, and t/a = 0.20, as the number of terms used in the field expansions in region 1 (N), and region 2 (M) is varied. Note that convergence of higher order eigenvalues as a function of order of the number of terms is achieved even more rapidly than the dominant eigenvalue. That is generally the case with higher order eigenvalues because the deviation from the ridgeless case becomes smaller. In the actual computations of (3.47), the summations over N and M are terminated when

Table 3.1
Convergence of Normalized Even- and Odd-Mode Impedances
Septum length = $0.8146a/\pi$
Septum thickness = $0.01094a$
Frequency = $1.4478 f_c$, f_c = TE_{10} cutoff frequency in rectangular waveguide

Number of Modes	Normalized Even-Mode Impedance	Normalized Odd-Mode Impedance
20	0.5923	0.2662
40	0.5909	0.2657
60	0.5919	0.2661
80	0.5923	0.2663
100	0.5921	0.2662
120	0.5919	0.2661
140	0.5919	0.2662
160	0.5921	0.2662
180	0.5921	0.2662

Table 3.2
Convergence of Insertion Loss
Frequency = $1.4478 f_c$, f_c = TE_{10} cutoff frequency in rectangular waveguide
Insert thickness = $0.01094a$

Septum lengths	Resonator lengths
$d_1 = 0.1528a/\pi$	$l_1 = 2.2055\ a/\pi$
$d_2 = 0.8146a/\pi$	$l_2 = 2.2568a/\pi$
$d_3 = 0.9985a/\pi$	$l_3 = 2.2603a/\pi$

Number of Modes	Insertion Loss (dB)
20	0.9602×10^{-2}
40	0.6994×10^{-2}
60	0.8659×10^{-2}
80	0.9354×10^{-2}
100	0.8993×10^{-2}
120	0.8583×10^{-2}
140	0.8646×10^{-2}
160	0.8848×10^{-2}
180	0.8888×10^{-2}

Table 3.3
Variation of Dominant Eigenvalue With Number of Terms in Region 1 (N) and
Region 2 (M) Parameters: a = 12.7 mm, b = 10.16 mm, t = 2.54 mm, s = 2.794 mm

M	N	k_c (rad/mm)
5	1	0.1439
5	5	0.1454
5	10	0.1468
5	15	0.1482
5	20	0.1503
5	30	0.1628
10	1	0.1449
10	5	0.1462
20	5	0.1439
30	5	0.1438

Table 3.4
Variation of Higher Eigenvalue With Number of Terms in Region 1 (N) and Region 2 (M)
Parameters: a = 12.7 mm, b = 10.16 mm, t = 2.54 mm, s = 2.794 mm

M	N	k_c (rad/mm)
5	1	9.3642
5	5	9.3637
5	10	9.3119
5	15	9.3645
5	20	9.3645
5	30	9.3645
10	1	9.3639
10	5	9.3628
20	5	9.3629
30	5	9.3628

N = 5 and M = 30 with enough accuracy when examining the convergence of the variation of the eigenvalue as a function of N and M. For example, by comparing the eigenvalues of this proposed analysis with those of [15,17,18] for double-ridged waveguide (Table 3.5), it can be seen that all eigenvalues exhibit a good agreement.

Table 3.6 shows the convergence of the reflection coefficient magnitude of a metal septum with a length of $0.6996a/p$ and thickness of $0.0043745a$ placed between two different ridged waveguides, with gaps of 9.00 and 1.00 mm, respectively, as the number of the modes used in the fields expansions is increased. The

Table 3.5
Cutoff Eigenvalues k_c (rad/mm) of the Five Modes in a Double-Ridged Waveguide
Parameters: a = 12.7 mm, b = 10.16 mm, t = 2.54 mm, s = 2.794 mm

Mode	1	2	3	4	5
Present method	0.1438	0.3162	0.6190	0.6712	0.6972
Ref. [18]	0.1438	0.3155	0.6215	0.6707	0.6971
Ref. [19]	0.1437	0.3166	0.6190	0.6712	0.6973
Ref. [15]	0.1439	0.3166	0.6191	0.6711	0.6974

Table 3.6
Convergence of Module of Reflection Coefficients
Septum length = $0.6996a/\pi$
Septum thickness = $0.0043745a$
Frequency = $1.4453 f_c$, f_c = TE_{10} cutoff frequency in rectangular waveguide
Gap of the ridged waveguides (mm) = 9.00/1.00

Number of Modes	$|S_{11}|$
1	0.84375
5	0.97795
20	0.97907
40	0.97724
60	0.97928
80	0.97929
100	0.97928
120	0.97927
140	0.97926

convergence of the insertion loss of an asymmetrical fifth-degree ridged waveguide E-plane bandpass filter (nonuniform convergence) is presented in Table 3.7. As can be seen, the convergence of the reflection coefficient magnitude of the metal septum (Table 3.6) is very good. However, the convergence of the insertion loss of an asymmetrical fifth-degree ridged waveguide filter (Table 3.7) is not as good. Therefore, to ensure good accuracy, more than 140 modes need to be included in the field expansions. That is similar to the situation for the septum in rectangular waveguide and is due to the singular behavior of the magnetic field at the edges of the septum. The calculation of insertion loss requires the evaluation of the ABCD parameters of six E-plane septa. The same number of modes is used in all six cases. About 17 hours of CPU time on a SUN computer (SPARC station 10) are required for the calculation of the insertion loss of the filter at one frequency using 100

Table 3.7
Convergence of Insertion Loss
Insert thickness = 0.0043745a
Frequency = 1.4453f_c, f_c = TE$_{10}$ cutoff frequency in rectangular waveguide
Gap of the ridged waveguides (mm) = 9.00/1.00/8.00/2.00/7.00

Septum lengths	Resonator lengths
d_1 = 0.19643a/p	l_1 = 2.17314a/p
d_2 = 0.69969a/p	l_2 = 1.89891a/p
d_3 = 0.88734a/p	l_3 = 2.22128a/p
d_4 = 0.88467a/p	l_4 = 2.00274a/p
d_5 = 0.64905a/p	l_5 = 2.15683a/p
d_6 = 0.15275a/p	

Number of Modes	Insertion Loss (dB)
1	0.2261×10^{-1}
5	0.3590×10^{-1}
20	0.3748×10^{-1}
40	0.3985×10^{-1}
60	0.4564×10^{-1}
80	0.4621×10^{-1}
100	0.4508×10^{-1}
120	0.4364×10^{-1}
140	0.4239×10^{-1}

modes. Because the calculation of the insertion loss at one frequency using 100 modes requires about 17 hours of CPU time, the overall numerical effort can easily become exceedingly heavy. Therefore, working with a large number of modes places a heavy demand on computing resources.

3.3 Coplanar Waveguide Discontinuities

3.3.1 Introduction to Coplanar Waveguide Modeling

To design filters using CPWs [13,20] as the main transmission line, it is necessary to characterize as many standard subsections of the layout as possible. This section examines, through a variety of simulations and measurements, Em software, by Sonnet, which calculates S parameters for predominantly planar geometries using the method of moments. A wide range of electromagnetic simulators is available for CPW circuits. A list of some commercial electromagnetic simulators is given in Table 3.8. The listed programs are intended to solve for the S parameters of arbitrarily shaped CPW structures. By limiting the problem to predominantly planar

Table 3.8
Some Electromagnetic Simulators

Company	Product (all trademarks acknowledged)	Type
HP-EEsof (HP range)	Momentum	3D planar electromagnetic
	HFSS (HP/Ansoft)	3D arbitrary electromagnetic
Sonnet Software	Em	3D planar electromagnetic
Jansen Microwave	Unisym/Sfpmic	3D planar electromagnetic
Ansoft Corporation	Maxwell-Strata	3D planar electromagnetic
	Maxwell SI Eminence	3D arbitrary electromagnetic
ArguMens	Stingray	3D planar electromagnetic
Optimization System Associates	Empipe	Sonnet, Ansoft, HP optimization
Zeland Software	IE3D	3D arbitrary electromagnetic
MacNeal-Schwendler Corp.	MicroWaveLab	3D arbitrary electromagnetic
Kimberley Communications Consultants	Micro-Stripes	3D arbitrary electromagnetic (TLM)

structures, the speed of analysis is improved dramatically. Usually, the circuit conductors are divided into subsections and the method of moments is used for the electromagnetic analysis. An example of this type of simulator is Em. Often, many metal and dielectric layers can be handled, but these are assumed to be planar, so the term 2.5D has been coined (2D currents but 3D fields). These simulators cannot analyze true 3D structures, such as microstrip-to-stripline transitions, where the dielectrics are not planar. Yet when the current on the third dimension can be used to represent, for example, vias, but the analysis allows only layered dielectrics, the term 3D planar is more appropriate [21]. The speed of planar electromagnetic simulators makes them practical for carefully investigating nonstandard structures but conventional optimization is not really feasible. For all those numerical methods, the most time-consuming step is the solving of the matrix involving $O(n^3)$ number of operations, where n is the number of elements in the matrix, while the building of the matrix encompasses $O(n^2)$ operations. Recently, however, a fast iterative matrix solver in which solution time increases only as the square of the number of unknowns rather than the traditional cubic technique has been developed by Ansoft Corporation for its 3D planar Maxwell-Strata™ electromagnetic simulator. This method, coined ALPS, is an adaptive procedure that allows wideband S parameters data and radiation to be computed from a handful of single-frequency analyses at the dominant poles and zeros.

OSA (Canada), however, has made a breakthrough in electromagnetic optimization with the development of the Space-Mapping™ optimization technique

[22] used in the Empipe™ family. For 3D planar electromagnetic simulators, a minimum of 16 MB of RAM generally is required, and at least 32 MB of swap space is needed in the hard disk. Additionally, around 100 MB are needed to load and store the program

3.3.2 Introduction to Xgeom™ and Em™ Software

The remainder of this chapter briefly describes how the software packages Xgeom and Em can be used for modeling CPW discontinuities. Xgeom is an X-Window system mouse-based program used to capture the CPW input geometry to the electromagnetic analysis program Em. The output file is saved with the .geo extension. The software program Xgeom is for the precise analysis of passive microwave circuits. Xgeom provides a straightforward user interface that allows specification of all needed information concerning the circuit to be analyzed. After capturing the circuit with Xgeom and saving the resulting file, the user can run Em using the .geo file as input. Em automatically subsections the circuit and performs (all effects included) electromagnetic analysis. Em saves the resulting S parameters ready for input to major microwave design programs such as DBFILTER or Touchstone. Xgeom can capture (and Em can analyze) circuits with any number of dielectric layers with any thickness, dielectric constant, and loss tangent. Em is a 3D predominantly planar electromagnetic analysis tool that can calculate S parameters for planar geometries. It takes into account dispersion, stray coupling, discontinuities, surface waves, moding, metallization loss, dielectric loss, and radiation loss. The electromagnetic analysis of CPW or any other 3D planar geometry is done by solving the current distribution in the circuit metallization using the Method of Moments. It analyzes planar structures inside a shielding box. Em can calculate ultra-precise S parameters of specific discontinuities or groups of interacting discontinuities, and can grant design validation, thus eliminating the expensive design iterations of a circuit. Em can produce microwave package evaluation, providing an analysis of a package prior to fabrication. Also, a wide variety of CPW discontinuities can be evaluated using Em. Em also can handle metallization between any or all layers, including the effect of DC and skin effect loss as well as circuits with any number of ports. A new feature of this software is the push-pull (balanced, or odd-mode) ports that allows CPW structures to be analyzed. More information can be found in the *Em User's Manual* [12].

3.3.2.1 Analysis of CPW Discontinuities with *Xgeom* and *Em* Software

Circuits with any number of ports can be analyzed with *Em*, without limit, and the discontinuities between them can be evaluated precisely. That has an important application for filters, to calculate the discontinuity and incorporate it in circuit theory programs. The option is explained in detail in Section 12 of the *Em User's*

Manual[12]. CPW filters can be examined using Em by means of level-to-level vias to form air bridges and balanced ports. That is done in Xgeom by having negative labels in the ports, in the submenu *Objects/Ports/Renumber*. Em sums the total current going out of all the ports with the same negative port number. It is important that the ground lines touch the sidewall only at the location of negative port numbers. Balanced ports are also called push-pull ports.

Figure 3.13 shows the layout of a CPW transmission line element being displayed in Xgeom's window with push-pull ports (tabled −1, +1, −1, etc.) and arrows marking the reference planes, which are employed for de-embedding (de-embedding removes unwanted effects occurring at the box edges). Figure 3.13 also shows vias going between the upper metal layer and the lower metal underpass ground straps. Figure 3.14 shows the layout of the CPW discontinuities such as edge-coupled CPW lines (Figure 3.14a) and the CPW coupling gap (Figure 3.14b). That layout and the other layouts in this section were used as building blocks in CPW filter design, which is described in detail in Chapter 8. The CPW

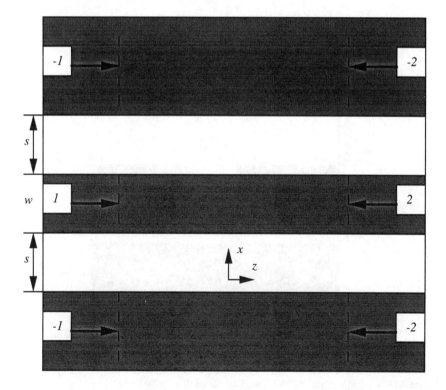

Figure 3.13 Drawing of the CPW transmission line (CPW resonator) subsection being displayed in *Xgeom*'s window with push-pull ports and reference planes that are used for de-embedding.

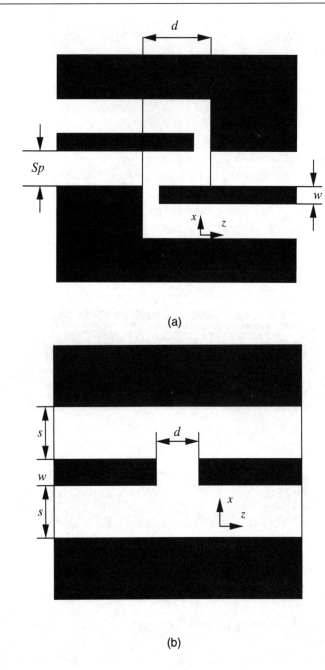

Figure 3.14 (a) Edge-coupled CPW lines and (b) CPW coupling gap.

line, which has a thin metal underpass connecting both coplanar ground planes, was characterized using an Em simulator. The analysis of this structure is well within the capability of most workstations, with a run time of a few minutes per frequency point. It should be pointed out, however, that there is a practical limit to the complexity of a layout that can be analyzed. Em is both a memory-intensive and a computationally intensive program. Small circuits are analyzed quickly, but the computing time and memory required to analyze a circuit increases dramatically with the number of subsections in a layout.

References

[1]   Rozzi, T., et al., "Accurate Full Band Equivalent Circuits of Inductive Posts in Rectangular Waveguide," *IEEE Trans. Microwave Theory & Tech.,* Vol. MTT-40, May 1992, pp. 1000–1009.

[2]   Tao, J. W., and H. Baudrand, "Multimodal Variational Analysis of Uniaxial Waveguide Discontinuities," *IEEE Trans. Microwave Theory & Tech.,* Vol. MTT-39, March 1991, pp. 506–516.

[3]   Konishi, Y., and H. Matsumura, "Short End Effect of Ridge Guide With Planar Circuit Mounted in Waveguide," *IEEE Trans. Microwave Theory & Tech.,* Vol. MTT-26, October 1978, pp. 716–719.

[4]   Ney, M. M., M. Chenier, and G. J. Costache, "Investigation on the Power Handling Capacity of a Glass of E-Plane Millimetre-Wave Filters Using Finite Element Modelling," *Int. J. Numerical Modelling,* Vol. 2, 1989, pp. 93–102.

[5]   Gebauer, A., and F. Hernandez-Gil, "Analysis and Design of Waveguide Multiplexers Using the Finite Element Method," *Proc. 18th European Microwave Conf.,* Stockholm, Sweden, 1988, pp. 521–524.

[6]   Itoh, T., *Numerical Techniques for Microwave and Millimeter-Wave Passive Structures,* New York: Wiley, 1989.

[7]   Wexler, A., "Solution of Waveguide Discontinuities by Modal Analysis," *IEEE Trans. Microwave Theory & Tech.,* Vol. MTT-15, September 1967, pp. 508–517.

[8]   Bornemann, J., "Comparison Between Different Formulations of the Transverse Resonance Field-Matching Technique for the Three Dimensional Analysis of Metal-Fined Waveguide Resonators," *Int. J. Numerical Modelling,* Vol. 4, March 1991, pp. 63–73.

[9]   Mansur, R. R., and R. H. Macphie, "An Improved Transmission Matrix Formulation of Cascaded Discontinuities and Its Application to E-Plane Circuits," *IEEE Trans. Microwave Theory & Tech.,* Vol. MTT-34, December 1974, pp. 1490–1498.

[10] Beyer, A., "Calculation of Discontinuities in Grounded Finlines Taking into Account the Metallization Thickness and the Influence of the Mount-Slits," *Proc. 12th European Microwave Conference,* Helsinki, Finland, 1982, pp. 681–686.

[11] Mittra, R., and S. W. Lee, *Analytical Techniques in the Theory of Guided Waves,* New York: Macmillan, 1971.

[12] *Em User's Manual, Vol. 1,* Release 4.0, Sonnet Software Inc., Liverpool, NY, 1996.

[13] Gupta, K. C., et al., *Microstrip Lines and Slotlines*, 2nd ed., Norwood, MA: Artech House, 1996.

[14] J. P. Montgomery, "On the Complete Eigenvalue Solution or Ridged Waveguide," *IEEE Trans. Microwave Theory & Tech.*, Vol. MTT-19, June 1971, pp. 547–555.

[15] Y. Utsumi, "Variation Analysis of Ridged Waveguide Modes," *IEEE Trans. Microwave Theory & Tech.*, Vol. MTT-33, February 1985, pp. 111–120.

[16] Collin, R. E., *Field Theory of Guided Waves*, New York: McGraw-Hill, 1960.

[17] Dasgupta, D., and P. K. Saha, "Eigenvalue Spectrum of Rectangular Waveguide With Two Symmetrically Placed Double Ridges," *IEEE Trans. Microwave Theory & Tech.*, Vol. MTT-29, January 1981, pp. 47–51.

[18] Hoefer, W. J. R., and M. R. Burtin, "Closed-Form Expressions for the Parameters of Finned and Ridged Waveguide," *IEEE Trans. Microwave Theory & Tech.*, Vol. MTT-30, December 1982, pp. 2190–2194.

[19] Pyle, J. R., "The Cutoff Wavelength of the TE10 Mode in Ridged Rectangular Waveguide of Any Aspect Ratio," *IEEE Trans. Microwave Theory & Tech.*, Vol. MTT-14, April 1966, pp. 175–183.

[20] Wen, C. P., "Coplanar Waveguide: A Surface Strip Transmission Line Suitable for Nonreciprocal Gyromagnetic Device Application," *IEEE Trans. Microwave Theory Tech.*, Vol. MTT-17, No. 12, December 1969, pp. 1087–1090.

[21] Rautio, J. C., "Some Comments on Electromagnetic Dimensionality," *IEEE Microwave Theory & Tech. Soc. Newsletter*, Winter 1992, pp. 23.

[22] Bandler, J. W., et al., "Electromagnetic Optimization Exploiting Aggressive Space Mapping," *IEEE Trans. on Microwave Theory & Tech.*, Vol. 43, December 1995, pp. 2874–2882.

Selected Bibliography

Bornemann, J., and F. Arndt, "Modal S-Matrix Design of Optimum Stepped Ridged and Finned Waveguide Transformers," *IEEE Trans. Microwave Theory & Tech.*, Vol. MTT-35, June 1987, pp. 561–567.

Bornemann, J., and F. Arndt, "Transverse Resonance, Standing Wave, and Resonator Formulations of the Ridge Waveguide Eigenvalue Problem and Its Application to the Design of E-Plane Finned Waveguide Filters," *IEEE Trans. Microwave Theory & Tech.*, Vol. MTT-38, August 1990, pp. 1104–1113.

Budimir, D., et al., "V-Shaped CPW Transmission Lines for Multilayer MMICs," *IEE Electronics Letters*, October 1995, pp. 1928–1929.

Chu, T. S., and T. Itoh, "Comparative Study of Mode-Matching Formulations for Microstrip Discontinuity Problems," *IEEE Trans. Microwave Theory & Tech.*, Vol. MTT-33, October 1985, pp. 1018–1023.

Cohn, S. B., "Properties of Ridge Waveguide," *Proc. IRE*, Vol. 35, August 1947, pp. 783–788.

Empire Reference Manual, Version 3.1, Optimization System Associates Inc., Canada, 1995.

Fan, P., and D. Fan, "Computer Aided Design of E-Plane Waveguide Passive Components," *IEEE Trans. Microwave Theory & Tech.*, Vol. MTT-37, February 1989, pp. 335–339.

Getsinger, W. J., "Ridged Waveguide Field Description and Application to Direction Couplers," *IRE Trans. Microwave Theory & Tech.*, Vol. MTT-10, January 1962, pp. 41–51.

HFSS Reference Manual, Release 2.0, Hewlett-Packard Co., Palo Alto, CA, 1992.

Hopfer, S., "The Design of Ridged Waveguide," *IRE Trans. Microwave Theory & Tech.,* Vol. MTT-3, October 1955, pp. 20–29.

Konishi, Y., and K. Uenakada, "The Design of a Band-Pass Filter With Inductive Strip-Planar Circuit Mounted in Waveguide," *IEEE Trans. Microwave Theory & Tech.,* Vol. MTT-22, October 1974, pp. 869–873.

Kuhn, E., "A Mode-Matching Method for Solving Field Problems in Waveguide and Resonator Circuits," *AEU,* Band 27, Heft 12, 1973, pp. 511–518.

LINMIC+ User Manual, Version 2.1, Jansen Microwave, Germany, 1989.

Mansur, R. R., R. S. K. Tong, and R. H. Macphie, "Simplified Description of the Field Distribution in Finlines and Ridge Waveguide and Its Application to the Analysis of E-Plane Discontinuities," *IEEE Trans. Microwave Theory & Tech.,* Vol. MTT-36, December 1988, pp. 1825–1832.

Omar, A. S., and K. Schunemann, "Transmission Matrix Representation of Finline Discontinuities," *IEEE Trans. Microwave Theory & Tech.,* Vol. MTT-33, September 1985, pp. 765–770.

Omar, A. S., and K. Schunemann, "Application of the Generalized Spectral-Domain Technique to the Analysis of Rectangular Waveguides With Rectangular and Circular Metal Inserts," *IEEE Trans. Microwave Theory & Tech.,* Vol. MTT-39, June 1991, pp. 944–952.

Patzelt, H., and F. Arndt, "Double-Plane Steps in Rectangular Waveguide and Their Application for Transformers, Irises and Filters," *IEEE Trans. Microwave Theory & Tech.,* Vol. MTT-30, May 1982, pp. 771–776.

Ramo, S., J. R. Whinnery, and T. VanDuzer, *Fields and Waves in Communication Electronics,* New York: Wiley, 1984.

Schwinger, J., and D. Saxon, *Discontinuities in Wave Guide (Documents on Modern Physics),* New York: Gordon and Greach, 1968.

Shih, Y. C., and K. Gray, "Convergence of Numerical Solutions of Step-Type Waveguide Discontinuity Problems by Modal Analysis," *IEEE MTT-S Int. Microwave Symp. Dig.,* 1983, pp. 233–235.

Shih, Y. C., and T. Itoh, "E-Plane Filters With Finite-Thickness Septa," *IEEE Trans. Microwave Theory & Tech.,* Vol. MTT-31, December 1983, pp. 1009–1013.

Shih, Y. C., T. Itoh, and L. Q. Bui, "Computer-Aid Design of Millimeter-Wave E-Plane Filters," *IEEE Trans. Microwave Theory & Tech.,* Vol. MTT-31, February 1983, pp. 135–142.

Si-fan, L., and C. Yi-yuan, "CAD of Rectangular and Ridged Waveguide Bandpass Filters," *IEEE MTT-S Int. Microwave Symp. Dig.,* 1982, pp. 532–534.

Tao, J. W., and H. Baudrand, "Rigorous Analysis of Triple-Ridge Waveguides," *IEE Electronic Letters,* Vol. 24, No. 13, June 1988, pp. 820–821.

Vahldieck, R., et al., "Optimized Waveguide E-Plane Metal Insert Filters for Millimeter-Wave Application," *IEEE Trans. Microwave Theory & Tech.,* Vol. MTT-31, January 1983, pp. 65–69.

Appendix 3A

Following Wexler's procedure [1], we enforce continuity of the transverse electric field at junction ($z = 0$) by equating (3A.1) with (3A.2), taking the cross-product with h_{am} and integrating, keeping in mind the orthogonality relation [2, p. 230]

$$\int_a e_{ai} \cdot h_{am} e_z ds = 0 \qquad (3A.1)$$

when $I \neq m$. The surface integral extends over the entire cross-section of the region a. The results are

$$(1 + \rho) a_1 \int_a e_{a1} \cdot h_{a1} e_z ds = \sum_{j=1}^{N} b_j \left[\int_b e_{bj} \cdot h_{a1} e_z ds \right.$$
$$\left. + \sum_{k=1}^{N} s_{jk} \int_b e_{bk} \cdot h_{a1} e_z ds \right] \qquad (3A.2)$$

for $m = 1$ and

$$a_m \int_a e_{am} \cdot h_{am} e_z ds = \sum_{j=1}^{N} b_j \left[\int_b e_{bj} \cdot h_{am} e_z ds \right.$$
$$\left. + \sum_{k=1}^{N} s_{jk} \int_b e_{bk} \cdot h_{am} e_z ds \right] \qquad (3A.3)$$

for $m \neq 1$.

To provide continuity of the transverse magnetic field through the aperture, we equate (3.1b) and (3.2b), take a cross-product with e_{bn}, and integrate. By using the orthogonality relation

$$\int_b e_{bn} \cdot h_{bj} e_z ds = 0 \qquad (3A.4)$$

when $n \neq j$, the result is

$$(1 - \rho) a_1 \int_b e_{bn} \cdot h_{a1} e_z ds - \sum_{i=2}^{M} a_i \int_b e_{bn} \cdot h_{ai} e_z ds =$$
$$\left(b_n - \sum_{j=1}^{N} b_j s_{jn} \right) \int_b e_{bn} \cdot h_{bn} e_z ds \qquad (3A.5)$$

If we now substitute (3A.3) into (3A.5) and change m to i, the a_i coefficients are eliminated, and we obtain the equation

$$\rho \int_b e_{bn} \cdot h_{a1}e_z ds + \sum_{j=1}^{N} \frac{b_j}{a_1} \sum_{i=2}^{M} \frac{\int_b e_{bj} \cdot h_{a1}e_z ds + \sum_{k=1}^{N} s_{jk} \int_b e_{bk} \cdot h_{a1}e_z ds}{\int_b e_{ai} \cdot h_{ai}e_z ds}$$

$$\int_b e_{bn} \cdot h_{ai}e_z ds + \left(\frac{b_n}{a_1}\right)\int_b e_{bn} \cdot h_{bn}e_z ds = \int_b e_{bn} \cdot h_{a1}e_z ds \tag{3A.6}$$

Equation (3A.6) is essentially N linear equations corresponding to $n = 1, 2, 3, \ldots, N$. There are $N+1$ unknowns, namely, ρ and the N modal coefficients in region b (b_1/a_1), (b_2/a_1), $\ldots$, (b_N/a_1), but by dividing (3A.2) by a_1 and rearranging, we have

$$\rho \int_b e_{bn} \cdot h_{a1}e_z ds - \sum_{j=1}^{N} \frac{b_j}{a_1}\left(\int_b e_{bj} \cdot h_{a1}e_z ds\right.$$

$$\left. + \sum_{k=1}^{N} s_{jk}\int_b e_{bk} \cdot h_{a1}e_z ds\right) = -\int_a e_{a1} \cdot h_{a1}e_z ds \tag{3A.7}$$

which, in combination with (3A.6), forms a system of $N+1$ linear equations with $N+1$ unknowns. The integrations are fairly straightforward for this problem and are as follows:

$$\int_a e_{ai} \cdot h_{ai}e_z ds = 0.5 a Y_{ai} \tag{3A.8}$$

$$\int_b e_{bj} \cdot h_{bj}e_z ds = 0.5 Y_{bj}(a - t) \tag{3A.9}$$

and

$$\int_b e_{bj} \cdot h_{ai}e_z ds = 0.5 Y_{ai}(a - t)\left[\frac{\sin(f)}{f}\right] \tag{3A.10}$$

where

$$f = \left[0.5p\left(\frac{a - t}{a}\right) - q \right] \pi \tag{3A.11}$$

and

$$g = \left[0.5p\left(\frac{a - t}{a}\right) + q \right] \pi \tag{3A.12}$$

The results in (3A.8) to (3A.12) are substituted into (3A.6) and (3A.7) when performing calculations. By solving the system of equations [i.e., (3A.6) and (3A.7)] for ρ, using both symmetrical and antisymmetrical excitations, equivalent T-circuit parameters (Figure 3.5) are determined. Assuming an open circuit at the central plane (Figure 3.4), $\rho = \rho_e$ can be computed from a system of $N + 1$ linear equations using $s_{jk} = +1$ and from it the normalized reactive even-mode impedance z_e is given by

$$z_e = \frac{1 + \rho_e}{1 - \rho_e} \tag{3A.13}$$

With the short circuit at the central plane (Figure 3.4), with $s_{jk} = -1$, the normalized reactive odd-mode impedance z_0 is given by

$$z_0 = \frac{1 + \rho_0}{1 - \rho_0} \tag{3A.14}$$

Appendix 3B: Scattering Parameters for Bifurcated Waveguides

Formulation 1A

$$S_{22} = (Y_a^{-1} GY_d + G)^{-1}(Y_a^{-1} GY_d - G)$$
$$S_{21} = 2(Y_a^{-1} GY_d + G)^{-1}$$
$$S_{12} = G(I + S_{22})$$
$$S_{11} = GS_{21} - I$$

Formulation 1B

$$S_{22} = (Y_a^{-1}GY_d + G)^{-1}(Y_a^{-1}GY_d - G)$$
$$S_{21} = 2(Y_a^{-1}GY_d + G)^{-1}$$
$$S_{12} = Y_a^{-1}GY_d(I - S_{22})$$
$$S_{11} = I - Y_a^{-1}GY_dS_{21}$$

Formulation 1C

$$S_{11} = (Y_d^{-1}G^tY_a + G^t)^{-1}(Y_d^{-1}G^tY_a - G^t)$$
$$S_{12} = 2(Y_d^{-1}G^tY_a + G^t)^{-1}$$
$$S_{21} = G^t(I + S_{11})$$
$$S_{22} = G^tS_{12} - I$$

Formulation 1D

$$S_{11} = (Y_d^{-1}G^tY_a + G^t)^{-1}(Y_d^{-1}G^tY_a - G^t)$$
$$S_{12} = 2(Y_d^{-1}G^tY_a + G^t)^{-1}$$
$$S_{21} = Y_d^{-1}GY_a(I - S_{11})$$
$$S_{22} = I - Y_d^{-1}GY_aS_{12}$$

Formulation 2A

$$S_{22} = (I + Y_d^{-1}G^tY_aG)^{-1}(I - Y_d^{-1}G^tY_aG)$$
$$S_{21} = 2(I + Y_d^{-1}G^tY_aG)^{-1}Y_d^{-1}G^tY_a$$
$$S_{12} = G(I + S_{22})$$
$$S_{11} = GS_{21} - I$$

Formulation 2B

$$S_{11} = (GY_d^{-1}G^tY_a + I)^{-1}(GY_d^{-1}G^tY_a - I)$$
$$S_{12} = 2(GY_d^{-1}G^tY_a + I)^{-1}G$$
$$S_{21} = Y_d^{-1}G^tY_a(I - S_{11})$$
$$S_{22} = I - Y_d^{-1}G^tY_aS_{12}$$

Formulation 2C

$$S_{22} = (G^t Y_a^{-1} GY_d + I)^{-1}(G^t Y_a^{-1} GY_d - I)$$

$$S_{21} = 2(G^t Y_a^{-1} GY_d + I)^{-1} G^t$$

$$S_{12} = Y_a^{-1} GY_d(I - S_{22})$$

$$S_{11} = I - Y_a^{-1} GY_d S_{21}$$

Formulation 2D

$$S_{11} = (I + Y_a^{-1} GY_d G^t)^{-1}(I - Y_a^{-1} GY_d G^t)$$

$$S_{12} = 2(I + Y_a^{-1} GY_d G^t)^{-1} Y_a^{-1} GY_d$$

$$S_{21} = G_t(I + S_{11})$$

$$S_{22} = G^t S_{12} - I$$

Appendix References

[1]   Wexler, A., "Solution of Waveguide Discontinuities by Modal Analysis," *IEEE Trans. Microwave Theory & Tech.,* Vol. MTT-15, September 1967, pp. 508–517.

[2]   Itoh, T., *Numerical Techniques for Microwave and Millimeter-Wave Passive Structures,* New York: Wiley, 1989.

4

Optimization-Based Filter Design

This chapter attempts to show how problems within the scope of filter design can be formulated effectively as optimization problems. The differences between optimizations are explained, and an appropriate method indicated. Optimization can be implemented in situations when the classical synthesis approach is inappropriate.

4.1 Filter Circuit Parameters

Figure 4.1 shows a two-port lossless reciprocal network as a filter. It operates between a resistive source and a resistive load. This network is characterized by its scattering matrix $[S(\omega)]$ (4.1a) normalized to the source impedance of 1Ω and the load impedance $R_L\Omega$.

$$[S(\omega)] = \begin{bmatrix} S_{11}(\omega) & S_{21}(\omega) \\ S_{21}(\omega) & S_{22}(\omega) \end{bmatrix} \tag{4.1a}$$

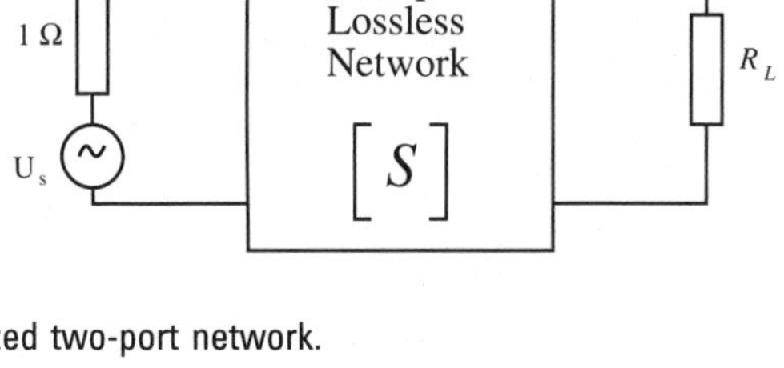

Figure 4.1 Normalized two-port network.

where

$$S_{ij}(\omega) = |S_{ij}(\omega)|e^{j\phi_{ij}(\omega)} \qquad (4.1b)$$

Some of the commonly used terms (see Figure 4.2) in filter design are defined here [1].

Center frequency (for bandpass filters) is defined as

$$f_0 = \sqrt{f_L f_H} \qquad (4.1c)$$

where f_L and f_H are the upper and lower bandedge frequencies, respectively.

Degree is defined as the number of elements or resonator sections used in a filter.

Bandwidth (in megahertz or percentage) is defined as the frequency difference between specified attenuation levels, generally 1 dB or 3 dB.

Insertion loss, L_I (in decibels), is defined as

$$L_I = 10 \, \log_{10}\left(\frac{1}{|S_{21}|^2}\right) \qquad (4.1d)$$

L_I is a measure of the total attenuation of a signal after the signal passes through the filter. It includes dissipation (or resistive) loss and the effects of input/output mismatch loss. In the case of a bandpass filter, insertion loss is proportional to the center frequency and degree and inversely proportional to bandwidth and the unloaded Q of the medium.

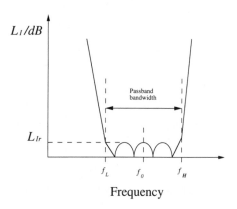

Figure 4.2 Insertion loss response of bandpass filter.

Ripple, L_{ir}, is defined as the variation of insertion loss amplitude across a defined frequency range.

Return loss, R_L (in decibels), is defined by

$$R_L = 10 \log_{10}\left(\frac{1}{|S_{11}|^2}\right) \tag{4.1e}$$

R_L is a measure of the input and output match to the characteristic impedance of the medium.

Selectivity is a measure of the rate of increase of attenuation at the band edge.

Rejection is the through attenuation to unwanted signals.

Group Delay, t_d (in seconds), is the time taken for an item of information to transmit from the input to the output of the device. Group delay is proportional to the rate of change of electrical phase, as shown by

$$t_d = -\frac{d\phi}{d\omega} \tag{4.1f}$$

where ϕ is the transmission phase (in radians), and ω is the angular frequency (in radians per second).

Power handling is the amount of RF power that can be handled by the filter without voltage breakdown. The power-handling level must be specified so the appropriate filter design is selected. Waveguide and coaxial filters have to be used for higher power applications, while printed circuit filters usually are used for lower power applications (up to a few hundred watts).

Temperature range is the ambient temperature range over which the filter must meet the full electrical specification.

4.2 Filter Design

Selection of the prototype characteristic function that most closely approximates the required filter characteristic is the first step of any filter design. Several transfer functions can be used. Figure 4.3 shows (a) Chebyshev, (b) generalized Chebyshev or quasi-elliptic, and (c) elliptic function responses. The second step is synthesis of a lowpass prototype network. Once the degree and type of transfer function required have been determined, the prototype network is mathematically transformed into lowpass, highpass, bandpass, and bandstop filters, either in distributed or lumped form. Sometimes the final filter can be synthesized directly using either distributed or lumped techniques.

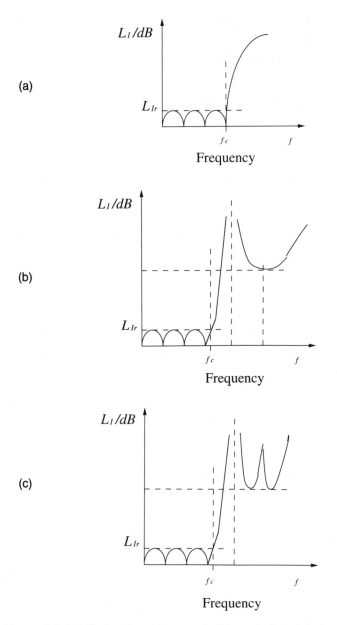

Figure 4.3 (a) Chebyshev function response; (b) quasi-elliptic function response; (c) elliptic function response.

The next step is the realization phase, in which the element values, that is, transmission lines, inductance, and capacitance are converted into realizable entities. The following parameters are the most important of any filter:

- Frequency;
- Bandwidth;
- Insertion loss;
- Selectivity;
- Rejection;
- Group Delay;
- Mechanical consideration.

The desired frequency and bandwidth of filter tend to dictate the medium in which a filter is best fabricated. Some applications and media are described in Chapter 2.

Classification of filter networks is shown in Figure 4.4. Table 4.1 shows characteristics of various filter realization media [2].

4.3 Filter Optimization

When a common approach to the design of filters (initial filter design in Figure 4.5) results in a design passband that differs considerably from that which is specified, optimization is required to tune the filter dimensions to achieve a design that meets certain requirements. Most microwave filters have not yielded exact optimum synthesis. Taking into account parasitic effects, high-frequency operation, frequency-dependent elements, a narrow range of element values, and so on, a common approach to design provides, at best, only approximate answers. Not infrequently, a common approach can be used to great advantage in providing the initial points for optimization.

This book introduces an optimization procedure based on Cohn's equal-ripple optimization to optimize filters based on Chebyshev function prototype. This method searches for tuning points in the filter transfer function and forces the ripple levels at those points to have specified values. The method requires knowledge of the filter insertion or return loss at those points. The method generates a set of equations that are solved to give a new set of parameter values. The cycle then is repeated until the filter characteristic is within an arbitrarily close value to the desired specification. The technique requires less calculation of the electrical parameters of filter discontinuity than generalized

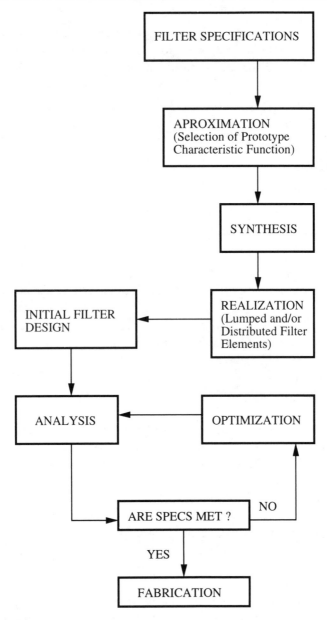

Figure 4.4 Classification of filters.

optimization routines so far applied. Cohn's technique optimizes the passband of a filter with respect to the Chebyshev (or minimax) criterion. The Chebyshev criterion can be defined as follows [3]:

An approximating function $F(x)$ constitutes an optimal Chebyshev approximation of a target function $f(x)$, on an interval [a,b], when

Table 4.1
Characteristics of Various Filter Realization Media

Media	Frequency (GHz)	Bandwidth (%)	Q (at 10 GHz)
Coaxial	0.10–40.00	1.0–30	2,000
Waveguide	1.00–100.0	0.1–20	5,000
Stripline	0.10–20.00	5-octave	150
Microstrip	0.10–100.0	5-octave	200
Suspended microstrip	1.00–200.0	2.0–20.0	1,000
Finline	20.0–200.0	2.0–50.0	500
Lumped elements	0.01–10.00 (hybrid)	20-octave	200
	0.10–60.0 (monolithic)	20-octave	100
Dielectric resonator	0.9–40.00	0.2–20	10,000

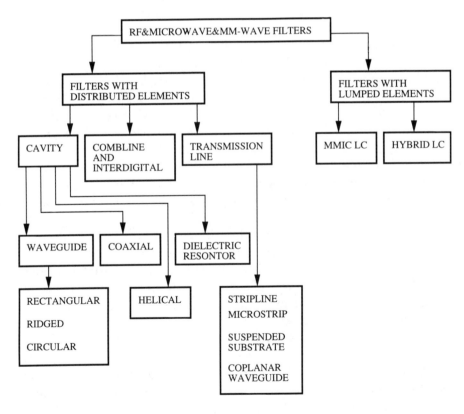

Figure 4.5 Flow chart for direct search optimization method.

$$E = \max|F(x) - f(x)| \quad x \subset [a, b] \qquad (4.1\text{g})$$

is minimal with respect to all the accepted functions $F(x)$. The variable x is real. Whether the requirements concern attenuation or phase response, it is usually required that the real characteristic does not depart from a target function by more than a maximum deviation within a certain interval.

Here, E is the error function, which relates directly to the way filters are specified in practice. The advantages of this method are that the problems of local minima are avoided, it requires fewer sampling points to achieve convergence than generalized error-minimization algorithms, and the Chebyshev criterion is satisfied. This method can handle symmetrical and asymmetrical bandpass [4–9], lowpass [10], and highpass filters based on Chebyshev, generalized Chebyshev, and elliptic function prototypes.

General purpose optimization techniques based on least pth, least-square, minimax, or worst-case objective functions use general forms of error-minimization algorithms [11], which simply force the filter transfer characteristic to be within specified constraints, whereas a filter must have a specified ripple characteristic, for example, the Chebyshev function. Usually the response of an optimizable filter is sampled at a number of equally spaced frequencies, and the error between that sampled response and the desired response is computed at each frequency. The optimization program, through an iterative process, reduces the error to a minimum, arriving at a final filter design in terms of the optimized filter parameters. These optimization techniques (such as: gradient, gradient minimax, least pth, minimax, quasi-newton, random, random minimax, random maximizer) cannot be guaranteed to satisfy filter specifications and may even converge to a local minimum.

Two steps in the optimization algorithm are fundamental: the determination of a search direction and the search for the minimum in that direction. The determination of the search direction is the most difficult part of optimization. There are two different ways of carrying out the determination of the search direction: gradient methods and direct search methods. Gradient methods use information about derivatives of the performance functions (with respect to designable parameters) for arriving at the modified set of parameters. Figure 4.6 shows a flow chart for that method. The direct search algorithms do not use gradient information, and parameter modification is carried out by searching for the optimum in a systematic manner. A flow chart for that optimization method is shown in Figure 4.7.

Cohn's optimization technique seeks to make an approximately equal-ripple circuit response function exactly equal ripple by iterative adjustment of

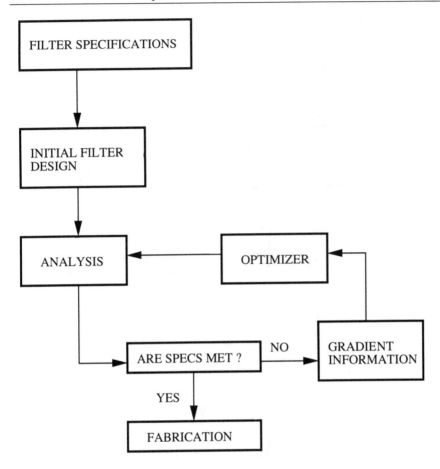

Figure 4.6 Flow chart for gradient optimization method.

the values of the circuit elements. Because the equal-ripple bandpass response for stepped impedance half-wave filters is optimum, in the sense of providing the minimum degree solution for a given passband and stop band specification [3], it seems appropriate to design filters by the application of Cohn's optimization technique to an approximate Chebyshev design. For a symmetrical filter with an equal ripple passband the characteristic function *g*, employing the terminology used by Gupta in his paper [12] on the design of multivariable lowpass equal-ripple filters, exhibits, within the passband, equal amplitude ripples that alternate in sign. The function

$$\rho' = \frac{g}{\sqrt{1 + g^2}} \tag{4.2}$$

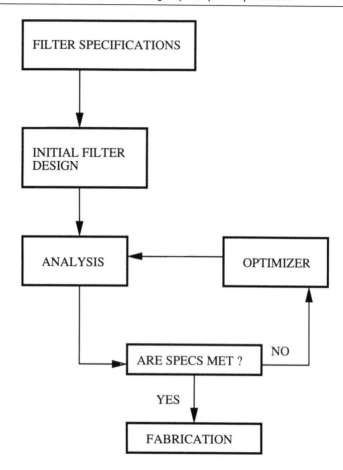

Figure 4.7 Flow chart for filter design.

also exhibits that property. Cohn assumed that for a sufficiently good approximate filter design g and ρ' would still exhibit ripples of alternating sign, but not necessarily equal amplitude. Cohn thus formulated the design of symmetrical filters by equal ripple optimization in terms of ρ', stating that g could just as well be used. Cohn stressed the importance of having a response function that exhibited ripples of alternate sign. ρ', as pointed out by Cohn, has the property that

$$|\rho'| = |S_{11}| \tag{4.3}$$

where S_{11} is the input reflection coefficient of the filter. For a symmetrical filter, $|S_{11}|$ will have the same number of ripples as ρ' and g, but the ripples will all be positive.

$|S_{11}|$ will, of course, be zero when ρ' and g are zero. Since in equal-ripple optimization it is necessary only to make the ripple amplitudes equal to some prescribed value, $|S_{11}|$ could just as well be employed as ρ' or g; it is not necessary to have a response function that will exhibit ripples of alternate sign. Any suitable function of $|S_{11}|$ also could be employed.

A similar equal-ripple optimization method has been described by Gupta for the design of multivariable lowpass equal ripple filters. Gupta's method [12] differs from Cohn's method as follows.

- Gupta's set of simultaneous equations includes equations to solve for the f_i (maxima frequencies) within the passband. Thus, the general bandpass filter with specified bandedges requires $2n$ simultaneous equations to be solved, rather than the $n + 1$ used in Cohn's method.

- He treats only lowpass symmetrical filters, which have $(n - 1)/2$ independent f_i, excluding the bandedge, and $(n + 1)/2$ independent parameters, requiring n simultaneous equations compared to $(n + 1)/2$ in Cohn's method.

- Gupta computes his response function from the ABCD matrix of the total filter, rather than that of its bisected half, as in Cohn's method.

By making the optimization algorithm force the zeros as well as the peaks of the equal-ripple error function, this method can be used for the asymmetrical case as well. If n peaks are present, an additional set of $n + 1$ sample functions will be needed for the zeros and the same number of additional equations will be required. Fortunately, the asymmetrical filter has the same number of additional optimization parameters.

Formulation of the equal-ripple optimization in the context of the design of symmetrical and asymmetrical filters, in terms of insertion loss, is given in Section 4.4. The numerical implementation of equal-ripple optimization, in the context of the design of lowpass, highpass, and bandpass filters based on the Chebyshev function prototype is presented in Chapters 5, 6, 7, and 8, respectively.

4.4 Description of the Algorithm

4.4.1 Symmetrical Case

To determine the degree of the filter (i.e., the number of resonators) required to satisfy a given bandpass filter specification, the equal-ripple bandpass characteristic proposed in [4] is frequently used in reactance coupled filters. This characteristic in terms of insertion loss, L_I, is given by

$$L_I = 10 \log_{10}\left\{1 + \epsilon^2 T_n^2\left[\frac{f_0}{f}\frac{\sin\left(\dfrac{\pi f}{f_0}\right)}{\pi\dfrac{(f_H - f_L)}{(f_H + f_L)}}\right]\right\} \qquad (4.4a)$$

where T_n is the nth degree Chebyshev polynomial of the first kind; ϵ defines the passband ripple level L_{ir}; f_H, and f_L are the upper and lower bandedge frequencies; and

$$f_0 = \sqrt{f_H f_L} \qquad (4.4b)$$

In general, approximate methods based on the synthesis of a Chebyshev prototype to the design of a symmetrical filter will not meet the specifications satisfied by (4.4). Assume that an mth degree symmetrical bandpass filter has an insertion loss response L_I of the form shown in Figure 4.8. It exhibits $m - 1(m = n + 1)$ zeros and $m - 2$ ripples, the maxima of which occur at the frequencies $f_2, f_3, \ldots, f_{m-}$.

For a symmetrical bandpass filter, all those $m - 2$ frequencies lie within the specified passband $f_l \Rightarrow f_h$. The deviation of a ripple maximum from the maximum allowed insertion loss in the passband, L_{Ir}, is a function of the $m = n + 1$ symmetrical filter parameter values required to specify the bandpass filter. There are $n - 1$ such functions for the symmetrical case:

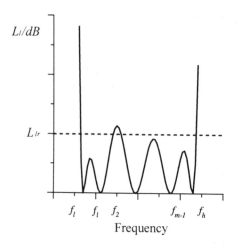

Figure 4.8 Scheme for the numerical optimization of symmetrical filter.

$$E_i = L_I(f_i) - L_{Ir} \quad i = 2, 3, \ldots, n \tag{4.5}$$

E_1 and E_m are defined by:

$$E_1 = L_I(f_l) - L_{Ir} \tag{4.6}$$

$$E_m = L_I(f_h) - L_{Ir} \tag{4.7}$$

E_1, E_m are also functions of the $m = n + 1$ parameter values of the symmetrical filter.

The specification

$$L_I(f) \le L_{Ir}, f_l \le f \le f_h \tag{4.8}$$

is satisfied when

$$E_i = 0 \quad i = 1, 2, 3, \ldots, m \tag{4.9}$$

This is a system of $m = n + 1$ nonlinear equations in $m = n + 1$ variables for the symmetrical case. Solving (4.9) gives the parameter values of a filter satisfying (4.8). The $E_i(i = 1, \ldots, m)$ can be regarded as the components of an m-dimensional error vector. Optimization is carried out by equating each of the components to zero (a vector process) rather than minimizing the magnitude of the vector (a scalar process). Thus, equal-ripple optimization can be regarded as a vector procedure, whereas general purpose optimization routines are scalar procedures. Usually the convergence criterion applied in general purpose optimization routines is that the gradient, with respect to the filter elements, of the magnitude of the error vector is zero. However, a zero gradient may correspond to a local minimum and the error may not be truly minimized.

The convergence criterion applied in equal-ripple optimization is that each component of the error vector is zero. Thus, on convergence the error is reduced to zero. The problem of local minima does not arise. Only the passband is optimized in equal-ripple optimization. Fixing the number and the amplitude of the ripples recovers the stop band performance of (4.4), except in the upper portion of the waveguide band, where, due to the frequency dependence of the electrical parameters of the E-plane septa, the assumptions made in [13] no longer are valid. In practice, (4.9) needs to be solved iteratively. To apply an iterative nonlinear equation solver in the case of symmetrical filters, it is necessary for a given set of filter parameter values to know the insertion loss only at the bandedge frequencies and at the ripple maxima. However, the frequencies at which the ripple maxima occur are unknown and are functions

of the filter parameter values. Those frequencies can be approximately located
by calculating the insertion loss on a coarse sample of frequency points in the
passband for a given set of filter parameter values. Figure 4.9 shows the ripple
maxima. Figure 4.9(a) shows f_i correctly centered at the maximum with

$$L_I(f_i - \Delta f), \ L_I(f_i + \Delta f) < L_I(f_i) \tag{4.10}$$

for maximum. In Figure 4.9(b), the sample frequency, f_i, is a little off, so the
function is sampled at frequencies slightly higher and lower.

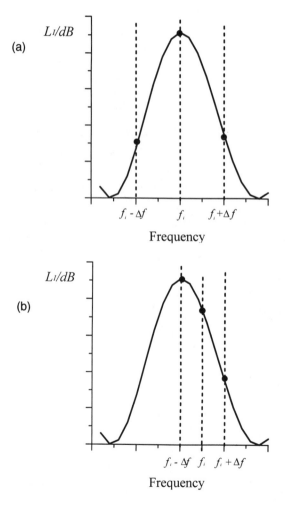

Figure 4.9 (a) f_i on maximum and (b) f_i off maximum.

By finding an equivalent parabola

$$L_I = a + bf + cf^2 \tag{4.11}$$

passing through the three points at $f_i - \Delta f$, f_i, and $f_i + \Delta f$, a correction is derived that can be applied to the frequency f_i to bring it closer to the extreme (maxima). The necessary condition for the maximum of $L_I(f)$ is that

$$\frac{dL_I}{df} = b + 2cf = 0 \tag{4.12}$$

that is,

$$f = -\frac{b}{2a} \tag{4.13}$$

where f locates the maximum of $L_I(f)$. The sufficiency condition for the maximum of $L_I(f)$ is that

$$\frac{d^2 L_I}{df^2} < 0 \tag{4.14}$$

The correct location and the amplitude of the ripple maxima can be found, by using the above procedure (quadratic interpolation [14]) in the last few iterations.

4.4.1.1 Algorithm for Solving System of Nonlinear Equations

The Newton-Raphson method [15] is a rapidly convergent technique for the solution of a system of nonlinear equations if a good initial approximation is available. The number of times the function is evaluated in the process of finding its root is the usual measure of computational effort. That includes function evaluations required to calculate derivatives numerically.

By using finite difference, the Jacobian matrix of the $n + 1$ nonlinear functions defined by (4.5) to (4.8) can be calculated numerically. For a given set of filter parameters, the finite difference calculation of the Jacobian matrix requires the evaluation of

$$E_i(x_1, x_2, \ldots, x_j + \delta x_j, \ldots, x_{n+1})$$
$$i = 1, 2, \ldots, n + 1,$$
$$j = 1, 2, \ldots, n + 1 \tag{4.15}$$

where $x_1, x_2, \ldots, x_{n+1}$ denote the $n + 1$ parameters required to specify a symmetrical filter.

By using sampling and quadratic interpolation, the evaluation of (E_i) generates as a by-product the value of

$$
\begin{gather}
f_i(x_1, x_2, \ldots, x_j + \delta x_j, \ldots, x_{n+1}) \\
i = 1, 2, \ldots, n - 1, \\
j = 1, 2, \ldots, n + 1
\end{gather}
\tag{4.16}
$$

The partial derivatives

$$
\frac{\partial f_i}{\partial x_j} \quad i = 1, 2, \ldots, n - 1, j = 1, 2, \ldots, n + 1
\tag{4.17}
$$

thus can be readily evaluated using finite difference. Denoting by x and E the $n + 1$ dimensional vectors with components $x_i(i = 1, 2, \ldots, n + 1)$ and $E_i(i = 1, \ldots, n + 1)$, the Newton-Raphson method has the general form [12]

$$
x^k = x^{k-1} - J^{-1}(x^{k-1})E(x^{k-1})
\tag{4.18}
$$

where k is the iteration number ($k = 1, 2, \ldots$) and J^{-1} is the inverse of the $m \times m$ Jacobian matrix evaluated at x^{k-1}. Once x^k has been calculated using (4.18), $f_i(x^k)(i = 1, 2, \ldots, n - 1)$ can be approximated by

$$
f_i(x^k) = f_i(x^{k-1}) + \sum_{j+1}^{n+1} \left(\frac{\partial f_i}{\partial x_j}\right)_{x=x^k} (x_j^k - x_j^{k-1})
\tag{4.19}
$$

Equation (4.19) identifies the regions within the passband that need to be sampled to calculate $E(x^k)$, as well as $J(x^k)$. The response and the errors after each iteration are computed again with the new corrected parameters, until the errors are judged to be sufficiently small. A subroutine, EROPTIM, has been developed and tested for several examples given in Chapters 5, 6, and 7.

4.4.1.2 Evaluation of the Jacobian

In the Newton-Raphson method [15], the most complex task is evaluating the Jacobian J of $E(x)$ in (4.9). That can be done either numerically or, when possible, analytically. In this work, a numerical approach has been adopted because we do not have analytic expressions for the functions. In general, an *m*-dimensional system of equations requires $m + 1$ function evaluations to calculate $E(x)$ and J numerically using finite difference. The Jacobian matrix of the m nonlinear functions is defined by

$$J = \begin{bmatrix} \dfrac{\partial E_1}{\partial y_1} & \dfrac{\partial E_1}{\partial y_2} & \cdots & \dfrac{\partial E_1}{\partial y_m} \\[2mm] \dfrac{\partial E_2}{\partial y_1} & \dfrac{\partial E_2}{\partial y_2} & \cdots & \dfrac{\partial E_2}{\partial y_m} \\[2mm] & & \vdots & \\[2mm] \dfrac{\partial f_m}{\partial y_1} & \dfrac{\partial f_m}{\partial y_2} & \cdots & \dfrac{\partial f_m}{\partial y_m} \end{bmatrix} \tag{4.20}$$

All blocks defined by (4.20) can be calculated numerically using finite difference for a given set of filter parameter values.

4.4.2 Asymmetrical Case

Figure 4.10 shows the form of insertion loss response L_I of an *n*th degree asymmetrical E-plane bandpass filter designed using approximate method. It exhibits $(m - 1)/2 (m = 2n + 1)$ nonzero minima, the minima of which occur at the frequencies $f_2, f_4, \ldots , f_{m-1}$. There are $(m - 3)/2$ ripples, the maxima of which occur at the frequencies $f_3, f_5, \ldots , f_{m-2}$. All those $m - 2$ frequencies lie within the specified passband $f_l \Rightarrow f_h$.

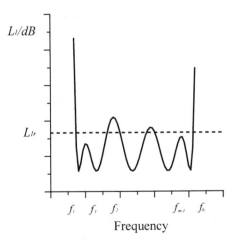

Figure 4.10 Scheme for the numerical optimization of asymmetrical filter.

The deviation of a ripple maximum from the maximum allowed insertion loss in the passband, L_{Ir}, is a function of the $m = 2n + 1$ values required to specify the bandpass filter. There are $2n - 1$ such functions

$$E_i = L_I(f_i) - L_{Ir} \quad i = 3, 5, \ldots, 2n - 1$$
$$E_i = L_I(f_i) \quad i = 2, 4, \ldots, 2n \qquad (4.21)$$

that have to be zero to satisfy the filter specification.

E_1 and E_m are defined by

$$E_1 = L_I(f_l) - L_{Ir} \qquad (4.22)$$

$$E_m = L_I(f_h) - L_{Ir} \qquad (4.23)$$

E_1, E_m are also functions of the $m = 2n + 1$ parameter values of the asymmetrical filter.

The specification

$$L_I(f) \le L_{Ir}, \, f_l \le f \le f_h \qquad (4.24)$$

$$L_I(f) = 0, \, f_l \le f \le f_h \qquad (4.25)$$

is satisfied when

$$E_i = 0 \quad i = 1, 2, 3, \ldots, m \qquad (4.26)$$

This is a system of $m = 2n + 1$ nonlinear equations in $m = 2n + 1$ variables that in practice needs to be solved iteratively. The parameter values of a filter satisfying (4.24) and (4.25) can be obtained by solving (4.26). The $E_i(i = 1, \ldots, m)$ can be regarded as the components of an m-dimensional error vector. By equating each of those components to zero (a vector process) rather than minimizing the magnitude of the vector (a scalar process), optimization is carried out.

To apply an iterative nonlinear equation solver, it is necessary for a given set of filter parameter values to know the insertion loss only at the bandedge frequencies and at the ripple maxima and minima. However, the frequencies at which the ripple maxima and minima occur are unknown and are functions of the filter parameter values. For a given set of filter parameter values, those frequencies can be approximately located by calculating the insertion loss on a coarse sample of frequency points in the passband. The ripple maxima and minima are shown in Figures 4.11 and 4.12, respectively.

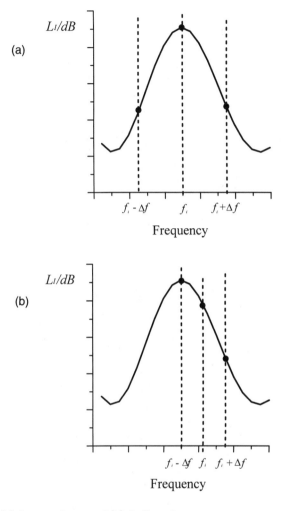

Figure 4.11 (a) f_i on maximum and (b) f_i off maximum.

Figures 4.11(a) and 4.11(b) show f_i correctly centered at the maximum and minimum with (4.9) for the maximum and

$$L_I(f_i - \Delta f),\ L_I(f_i + \Delta f) > L_I(f_i) \tag{4.27}$$

for the minimum.

In Figures 4.12(a) and 4.12(b), the sample frequency, f_i, is a little off, so the function is sampled at frequencies slightly higher and lower. By finding an equivalent parabola (4.11) passing through the three points at $f_i - \Delta f$, f_i, and $f_i + \Delta f$, a correction is derived that can be applied to the frequency f_i to

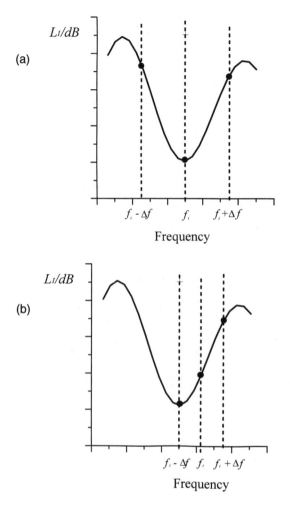

Figure 4.12 (a) f_i on maximum and (b) f_i off maximum.

bring it closer to the extreme (minima or maxima). The necessary condition for the maximum (or the minimum) of $L_I(f)$ is given by (4.12) and (4.13). The sufficiency condition for the maximum of $L_I(f)$ is given by (4.14) and

$$\frac{d^2 L_I}{df^2} > 0 \tag{4.28}$$

for the minimum.

By using that procedure (quadratic interpolation [14]) in the last few iterations, the correct location and amplitude of the ripple maxima and minima can be found.

In the asymmetrical case, the same algorithm can be used for solving the system of nonlinear equations as in the symmetrical case. The difference is only in the number of nonlinear functions. Instead of $n + 1$ nonlinear functions considered in the symmetrical case, in the asymmetrical case $2n + 1$ nonlinear functions defined by (4.21) to (4.25) will be considered. The Jacobian matrix of the $2n + 1$ nonlinear functions defined by (4.21) to (4.25) can be calculated numerically using finite difference, as in the symmetrical case.

References

[1] Filtronic Components Product Guide, Filtronic Components Ltd., England.

[2] Bahl, I. J., and P. Bhartia, *Microwave Solid State Circuit Design,* New York: Wiley, 1988.

[3] Hasler, M., and J. Neiryuck, *Electrical Filters,* Dedham, MA: Artech House, 1986.

[4] Cohn, S. B., "Synthesis of Commensurate Comb-Line Band-Pass Filters With Half-Length Capacitor Lines, and Comparison to Equal-Length and Lumped-Capacitor Cases," *IEEE MTT-S Int. Microwave Symp. Dig.,* May 1980, pp. 135–137.

[5] Hunton, J. K., "Novel Contributions to Microwave Circuit Design," *IEEE MTT-S Int. Microwave Symp. Dig.,* 1989, pp. 753–755.

[6] Postoyalko, V., and D. Budimir, "Design of Waveguide E-Plane Filters With All-Metal Inserts by Equal Ripple Optimization," *IEEE Trans. Microwave Theory and Techniques,* Vol. 42, No. 2, February 1994, pp. 217–222.

[7] Budimir, D., and V. Postoyalko, "EPFILTER: A CAD Package for E-Plane Filters," *Microwave J.,* August 1996, pp. 110–114.

[8] Budimir, D., "Optimized E-Plane Bandpass Filters With Improved Stop Band Performance," *IEEE Trans. Microwave Theory & Tech.,* February 1997, pp. 212–220.

[9] Parry, R., "Optimisation of Microwave Filters," *ESA Workshop on Advanced CAD for Microwave Filters and Passive Devices,* ESTEC/XRM, The Netherlands, November 6–8, 1995, pp. 265–271.

[10] Parry, R., "Optimisation of Microwave Filters," *Colloquium on Filters in RF and Microwave Communications,* Digest No.: 1992/220, University of Bradford, Bradford, England, December 1992, pp. 7/1–7/5.

[11] Bandler, J. W., "Optimization Methods for Computer-Aided Design," *IEEE Trans. Microwave Theory & Tech.,* Vol. MTT-17, August 1969, pp. 533–552.

[12] Gupta, G. P., "A Numerical Algorithm to Design Multivariable Low-Pass Equal-Ripple Filters," *IEEE Trans. Circuit Theory,* Vol. CT-20, 1973, pp. 161–164.

[13] Rhodes, J. D., "Microwave Circuit Realizations," in *Microwave Solid State Devices and Applications,* D. V. Morgan and M. J. Howes, eds., England: Peregrinus, 1980, pp. 49–57.

[14]   Gupta, K. C., R. Garg, and R. Chadha, *CAD of Microwave Circuits*, Dedham, MA: Artech House, 1981.

[15]   Ortega, J. M., and W. C. Rheinboldt, *Iterative Solution of Nonlinear Equations in Several Variables*. New York: Academic Press, 1970.

Selected Bibliography

Bandler, J. W., "Computer-Aided Circuit Optimization," in *Modern Filter Theory and Design*, G. C. Temes and S. K. Mitra, eds., New York: Wiley, 1973, pp. 211–271.

Bandler, J. W., and S. H. Chen, "Circuit Optimization: The State of the Art," *IEEE Trans. Microwave Theory & Tech.*, Vol. MTT-36, February 1988, pp. 424–443.

Bandler, J. W., et al., "Efficient Optimization With Integrated Gradient Approximations," *IEEE Trans. Microwave Theory & Tech.*, Vol. MTT-36, February 1988, pp. 444–455.

Broyden, C. G., "A Class of Methods for Solving Nonlinear Simultaneous Equations," *Mathematics of Computation*, Vol. 19, 1965, pp. 577–593.

Cohn, S. B., "Generalized Design of Bandpass and Other Filters by Computer Optimization," *IEEE MTT-S Int. Microwave Symp. Dig.*, June 1974, pp. 272–274.

DBFILTER Reference Manual, Tesla Communications Ltd., London, England.

FILTER Reference Manual, Eagleware Corp., USA, 1993.

Levy, R., "Theory of Direct Coupled Cavity Filters," *IEEE Trans. Microwave Theory & Tech.*, Vol. MTT-15, June 1967, pp. 340–348.

Lim, J. B., C. W. Lee, and T. Itoh, "An Accurate CAD Algorithm for E-Plane Type Bandpass Filters Using a New Passband Correction Method Combined With the Synthesis Procedures," *IEEE MTT-S Int. Microwave Symp. Dig.*, June 1990, pp. 1179–1182.

LINMIC+ Reference Manual, Jansen Microwave, Germany, 1989.

Matthaei, G., L. Young, and E. M. T. Jones, *Microwave Filters, Impedance-Matching Networks and Coupling Structures*, Dedham, MA: Artech House, 1980.

MDS Reference Manual, Hewlett-Packard Co., USA, 1994.

MMICAD Reference Manual, Optotek Ltd., Canada, 1995.

OSA90/HOPE Reference Manual, Optimization System Associates Inc., Canada, 1995.

SERIES IV/PC Reference Manual, Hewlett-Packard Co., USA, 1995.

Shih, Y. C., "Design of Waveguide E-Plane Filters With All Metal Inserts," *IEEE Trans. Microwave Theory & Tech.*, Vol. MTT-32, July 1984, pp. 695–704.

SUPER COMPACT Reference Manual, Compact Software Inc., USA, 1994.

Temes, G. C., and D. A. Calahan, "Computer-Aided Network Optimization the State of the Art," *Proc. IEEE*, Vol. 55, 1967, pp. 1832–1863.

TOUCHSTONE Reference Manual, EEsof Inc., USA, 1991.

5

Design of Lumped-Element Filters by Optimization

Because small size is an important parameter in some communication systems, lumped-element filters should be employed whenever possible. Besides size, they offer some other advantages over distributed filters, such as lower cost and broad stopbands free of spurious responses. Lumped-element filters constructed using air-wound inductors soldered into a small housing [1,2] and parallel plate chip capacitors, conventional thin-film techniques [3], thin-film/HTS techniques [4–6], and micromachining techniques [7] have been reported in the literature. These filters have been designed by approximate methods [8,9]. Results have been usually good, but errors can be quite large in broadband filters, and optimization is then required to tune the filter dimensions to satisfy the design specification.

Section 5.1 describes an approximate synthesis-based design procedure of lumped-element lowpass, highpass, bandpass, and bandstop filters. The numerical implementation of equal-ripple optimization, in the context of the design of lumped-element lowpass filters, is presented in Section 5.2. A filter that employs five elements with three inductors and two capacitors, or two inductors and three capacitors, with a cut-off frequency of 2 GHz is presented as a design example in Section 5.3.

5.1 An Approximate Synthesis-Based Design Procedure

This section examines a design approach for passive lumped-element filters. Figure 5.1 shows a two-port lossless reciprocal network, which operates between

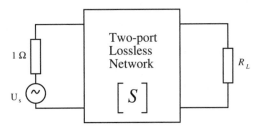

Figure 5.1 Normalized two-port lossless network.

a resistive source and a resistive load, as the simplest type of filter. The two-port lossless reciprocal network is characterized by its scattering matrix $[S(\omega)]$, as shown in (5.1a), normalized to the source impedance of $1\,\Omega$ and the load impedance of $R_L\Omega$.

$$[S(\omega)] = \begin{bmatrix} S_{11}(\omega) & S_{21}(\omega) \\ S_{21}(\omega) & S_{22}(\omega) \end{bmatrix} \qquad (5.1a)$$

where

$$S_{ij}(\omega) = |S_{ij}(\omega)|e^{j\phi_{ij}(\omega)} \qquad (5.1b)$$

$S_{21}(\omega)$ represents the scattering transmission coefficient of a two-port lossless network. In the complex frequency plane, the transfer function $S_{21}(s)$ can be defined as

$$S_{21}(s) = \frac{N(s)}{D(s)} \qquad (5.1c)$$

The realizability condition of a two-port lossless network constrains $N(s)$ to be a Hurwitz polynomial (i.e., all zeros of $N(s)$ must lie in the left s-half plane). Of course, $D(s)$ is a strictly Hurwitz polynomial.

5.1.1 Element Values of Chebyshev Lowpass Prototype Filters

A lowpass filter with $\omega' = 1$ as normalized cut-off frequency and $1\,\Omega$ as source and load resistors is defined as the lowpass filter prototype (Figure 5.2). For Chebyshev (equiripple passband and maximally flat stopband) response, the insertion loss (Figure 5.3) can be expressed as

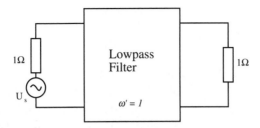

Figure 5.2 Lowpass filter prototype.

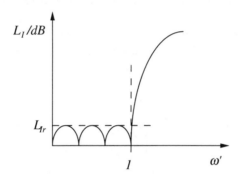

Figure 5.3 Chebyshev lowpass filter response.

$$L_I = 10 \log_{10}\left(\frac{1}{|S_{21}|^2}\right) \text{ [dB]} \tag{5.2}$$

with

$$|S_{21}|^2 = \frac{1}{1 + \epsilon^2 T_n^2(\omega')} \tag{5.3}$$

where $T_n(\omega')$ is the nth-degree Chebyshev polynomial of the first kind defined by

$$T_n(\omega') = \cosh[n \operatorname{Arcosh}(\omega')], \text{ for } |\omega'| \geq 1 \tag{5.4a}$$

$$T_n(\omega') = \cos[n \arccos(\omega')], \text{ for } 0 < \omega' \leq 1 \tag{5.4b}$$

and ϵ is the ripple level defined by

$$\epsilon^2 = 10^{(L_{Ir}/10)} - 1 \qquad (5.5)$$

where L_{Ir} is the passband ripple in decibels.

This response can be realized by a ladder network with inductors as series elements and capacitors as shunt elements. Several possible forms of a lowpass filter prototype are shown in Figure 5.4. The total number of elements required is given by the value of n in (5.2). The prototypes are easily changed to other impedances levels and frequency scales by using transformations applied to the filter elements.

The element values g_k (inductance if a series element and capacitance if a shunt element) for a Chebyshev lowpass prototype can be calculated from the following equations:

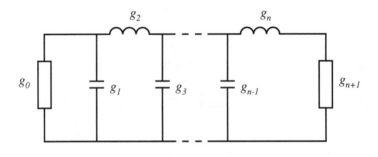

n even

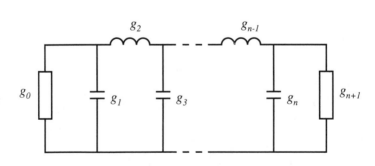

n odd

Figure 5.4 Several possible forms of a lowpass filter prototype.

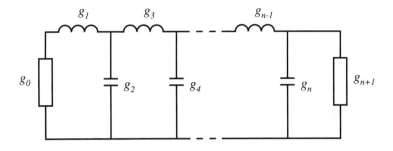

n even

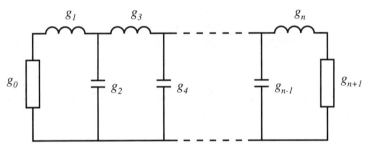

n odd

Figure 5.4 (continued).

$$g_0 = 1$$

$$g_1 = \frac{2 \sin\left(\dfrac{\pi}{2n}\right)}{\sinh\left(\dfrac{1}{n}\sinh^{-1}\dfrac{1}{\epsilon}\right)}$$

$$g_i g_{i+1} = \frac{4 \sin\left(\dfrac{2i-1}{2n}\pi\right) \sin\left(\dfrac{2i+1}{2n}\pi\right)}{\sinh^2\left(\dfrac{1}{n}\sinh^{-1}\dfrac{1}{\epsilon}\right) + \sin^2\left(\dfrac{i\pi}{n}\right)} \qquad i = 1, 2, \ldots, (n-1)$$

$$g_{n+1} = 1, \text{ for } n \text{ odd}$$

$$g_{n+1} = \frac{1}{\left(\epsilon + \sqrt{1 + \epsilon^2}\right)^2}, \text{ for } n \text{ even} \qquad (5.6)$$

5.1.2 Filter Design Procedure

A common approach to the design of the lumped-element lowpass, highpass, bandpass, and bandstop filters is well described in the literature [8,9]. Here, only the most important steps in the design procedure are presented.

5.1.2.1 Lowpass Filters

For a given filter specification such as the cutoff frequency yielding ω_c, passband return loss (L_R), stopband attenuation (L_I) at frequency f_s, the design procedure is summarized as follows:

1. Determine the passband ripple level ϵ from the minimum passband return loss, which is defined as:

$$L_R = 10 \log_{10}\left(1 + \frac{1}{\epsilon^2}\right) \qquad (5.7)$$

2. Determine the number of elements n at the designated stopband frequency f_s from (5.2) where application of the frequency transformation

$$\omega' = \frac{\omega}{\omega_c} \qquad (5.8)$$

that will transform the cut-off frequency from $\omega' = 1$ to $\omega' = w_c$ will be used.

3. Determine the prototype element values g_i from (5.6).

4. Determine the lumped-element values.

To determine the lumped-element values, the transformations of lowpass prototype into lowpass filter with arbitrary source and load impedances, Z_0 (Figure 5.5), are used.

These values are given by

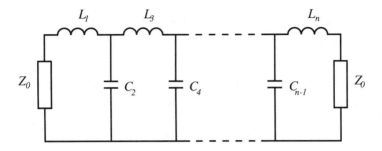

Figure 5.5 Lumped-element lowpass filter.

$$L_i = g_i\left(\frac{Z_0}{\omega_c}\right) \qquad (5.9)$$

for series inductors, and

$$C_i = g_i\left(\frac{1}{Z_0\omega_c}\right) \qquad (5.10)$$

for shunt capacitors.

5.1.2.2 Highpass Filters

For a given filter specification such as the cutoff frequency yielding ω_c, passband return loss (L_R), stopband attenuation (L_I) at frequency f_s, the design procedure is summarized as follows:

1. Determine the passband ripple level ϵ from (5.7).
2. Determine the number of elements n at the designated stopband frequency f_s from (5.2) where application of the frequency transformation

$$\omega' = -\frac{\omega_c}{\omega} \qquad (5.11)$$

that will transform the lowpass prototype with cutoff frequency of $\omega' = 1$ into a highpass filter with a cut-off frequency of $\omega' = w_c$ will be used.

3. Determine the prototype element values g_i from (5.6).

4. Determine the lumped-element values.

To determine the lumped-element values the transformations of lowpass prototype into highpass filter with arbitrary source and load impedances, Z_0 (Figure 5.6), are used.

Those values (Figure 5.7) are given by

$$C_i = \frac{1}{g_i Z_0 \omega_c} \tag{5.12}$$

for series capacitors, and by

$$L_i = \frac{Z_0}{g_i \omega_c} \tag{5.13}$$

for shunt inductors.

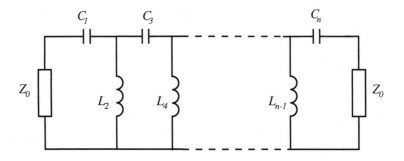

Figure 5.6 Lumped-element highpass filter.

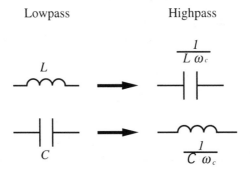

Figure 5.7 Lowpass-to-highpass transformation of filter elements.

5.1.2.3 Bandpass Filters

For a given filter specification such as the two passband edge frequencies yielding ω_L and ω_H, passband return loss (L_R), stopband attenuation (L_I) at frequency f_s, the design procedure is summarized as follows:

1. Determine the passband ripple level ϵ from (5.7).
2. Determine the number of elements n at the designated stopband frequency f_s from (5.2) where application of the frequency transformation

$$\omega' = \frac{1}{\delta}\left(\frac{\omega}{\omega_0} - \frac{\omega_0}{\omega}\right) \tag{5.14}$$

with

$$\delta = \frac{\omega_H - \omega_L}{\omega_0} \tag{5.15}$$

and

$$\omega_0 = \sqrt{\omega_H \omega_L} \tag{5.16}$$

that will transform the lowpass prototype with cutoff frequency of $\omega' = 1$ into a bandpass filter with the two passband edge frequencies yielding ω_L and ω_H, will be used.

3. Determine the prototype element values g_i from (5.6).
4. Determine the lumped-element values.

To determine the lumped-element values, the transformations of lowpass prototype into bandpass filter with arbitrary source and load impedances, Z_0 (Figure 5.8), are used.

Those values (Figure 5.9) are given by

$$L_i = g_i \frac{Z_0}{2\pi\Delta\omega}; \quad C_i = \frac{2\pi\Delta\omega}{g_i Z_0 \omega_0^2} \tag{5.17}$$

for series-tuned series elements, and by

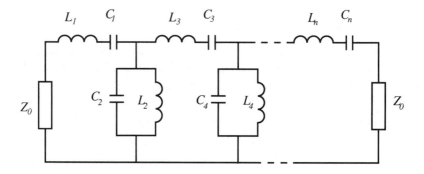

Figure 5.8 Resonator ladder lumped-element bandpass filter.

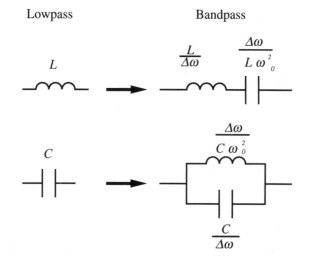

Figure 5.9 Lowpass-to-bandpass transformation of filter elements.

$$L_i = \frac{2\pi Z_0 \Delta\omega}{g_i \omega_0^2}; \quad C_i = \frac{g_i}{2\pi_i Z_0 \Delta\omega} \tag{5.18}$$

for shunt-tuned shunt elements, where

$$\Delta\omega = \omega_H - \omega_L \tag{5.19}$$

and

$$\omega_0 = \sqrt{\omega_H \omega_L} \tag{5.20}$$

5.1.2.4 Bandstop Filters

For a given filter specification such as the two bandstop edge frequencies yielding ω_L and ω_H, passband return loss (L_R), bandpass attenuation (L_I) at frequency f_s, the design procedure is summarized as follows:

1. Determine the passband ripple level ϵ from (5.7).
2. Determine the number of elements n at the designated bandpass frequency f_s from (5.2) where application of the frequency transformation

$$\omega' = \frac{1}{\dfrac{1}{\delta}\left(\dfrac{\omega}{\omega_0} - \dfrac{\omega_0}{\omega}\right)} \qquad (5.21)$$

   with

$$\delta = \frac{\omega_H - \omega_L}{\omega_0} \qquad (5.22a)$$

   and

$$\omega_0 = \sqrt{\omega_H \omega_L} \qquad (5.22b)$$

   that will transform the lowpass prototype with cutoff frequency of $\omega' = 1$ into a bandstop filter with the two bandstop edge frequencies yielding ω_L and ω_H, will be used.
3. Determine the prototype element values g_i from (5.6).
4. Determine the lumped-element values. To determine the lumped-element values, the transformations of lowpass prototype into bandstop filter (Figure 5.10a) with arbitrary source and load impedances, Z_0, are used.

   Those values (Figure 5.10b) are given by

$$C_i = \frac{1}{2\pi Z_0 g_i \Delta\omega}; \; L_i = \frac{2\pi Z_0 g_i \Delta\omega}{\omega_0^2} \qquad (5.23)$$

   for series-tuned series elements and by

$$C_i = \frac{2\pi g_i \Delta\omega}{Z_0 \omega_0^2}; \; L_i = \frac{Z_0}{2\pi_i g_i \Delta\omega} \qquad (5.24)$$

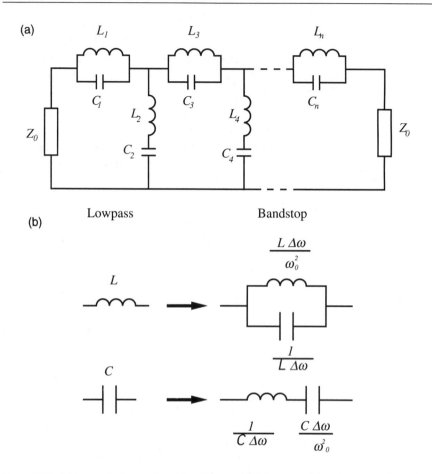

Figure 5.10 (a) Lumped-element bandstop filter and (b) lowpass-to-bandstop transformation of filter elements.

for series-tuned shunt elements, with

$$\Delta\omega = \omega_H - \omega_L \qquad (5.25)$$

and

$$\omega_0 = \sqrt{\omega_H \omega_L} \qquad (5.26)$$

5.1.2.5 Immittance Inverters

The design procedure of the bandpass lumped-element filters described in the preceding sections is useful for broadband design. However, if a narrowband bandpass filter is required, then an alternative design procedure is more suitable.

Such a design procedure includes the concept of immittance inverters with one kind of lumped element. The impedance inverter is defined by its ABCD matrix

$$[ABCD] = \begin{bmatrix} 0 & jK \\ j/K & 0 \end{bmatrix}$$ (5.27)

The function of the inverter is illustrated in Figure 5.11. Terminating the inverter by the load impedance (Z_L) or admittance (Y_L), the input impedance (Z_{in}) or input admittance (Y_{in}) of the inverter is given by

$$Z_{in} = \frac{K^2}{Z_L}$$ (5.28)

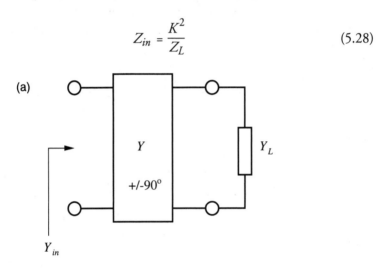

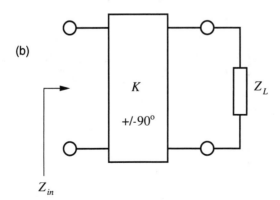

Figure 5.11 Inverters: (a) admittance inverter and (b) impedance inverter.

for K inverter, and by

$$Y_{in} = \frac{J^2}{Y_L}$$ (5.29)

for J inverter, where $K(Y)$ is the characteristic impedance (admittance) of the inverter.

For example, by terminating the inverter in an inductor of impedance Z_L, the input impedance of the inductively terminated inverter is the impedance of a capacitor. Thus, an LC ladder network (see Figure 5.5) is equivalent to a network consisting of series inductors separated by impedance inverters, as shown in Figure 5.12. The element values of this alternative prototype network are given by

$$L_I = \frac{2}{y} \sin\left[\frac{(2I - 1)\pi}{2n}\right] \quad I = 1, 2, 3, 4, \ldots, n$$ (5.30a)

and

$$K_{I,I+1} = \frac{\sqrt{y^2 + \sin(I\pi/n)}}{y} \quad I = 1, 2, 3, 4, \ldots, n-1$$ (5.30b)

with

$$y = \sinh\left[\frac{1}{n} \sinh^{-1}\left(\frac{1}{\epsilon}\right)\right]$$ (5.31)

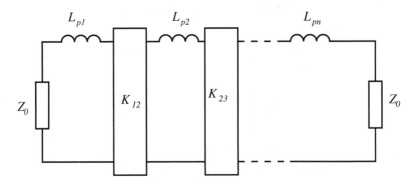

Figure 5.12 Lumped-element lowpass filter using impedance inverters.

Similarly, a bandpass filter can be realized by series inductors-capacitors (*L-C*) and series resonant circuits separated by impedance inverters (Figure 5.13), or parallel inductors-capacitors (*L-C*) and parallel resonant circuits separated by admittance inverters (Figure 5.14). Thus, the use of immittance inverters for filter design is a practical approach, since microwave filters realized by alternate series and parallel resonators are difficult to realize. The element values of this modified prototype network using impedance inverters are given by

$$L_{pl}C_{pl} = \frac{1}{\sqrt{\omega_0}} \quad I = 1, 2, 3, 4, \ldots, n \quad (5.32a)$$

$$K_{01} = \sqrt{\frac{\omega_0 \delta L_{p1} Z_0}{g_1}} \quad (5.32b)$$

$$K_{I,I+1} = \omega_0 \delta \sqrt{\frac{L_{pl} L_{p,I+1}}{g_I g_{I+1}}} \quad I = 1, 2, 3, 4, \ldots, n - 1 \quad (5.32c)$$

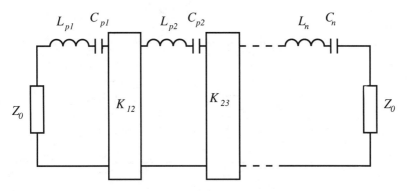

Figure 5.13 Lumped-element bandpass filter using impedance inverters.

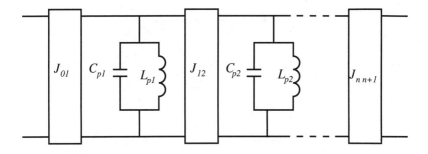

Figure 5.14 Lumped-element bandpass filter using admittance inverters.

$$K_{n,n+1} = \sqrt{\frac{\omega_0 \delta L_{pn} Z_0}{g_n}}$$ (5.32d)

with

$$\delta = \frac{\omega_H - \omega_L}{\omega_0}$$ (5.32e)

and

$$\omega_0 = \sqrt{\omega_H \omega_L}$$ (5.32f)

The prototype element values g_i are obtained from (5.6), while L_{pi}, Z_0, and C_{pi} can be chosen as desired. Four types of circuits that can be used as K or J inverters for the filter design are shown in Figure 5.15. Once the elements of the filter have been found, the frequency response of the overall filter at each frequency can be simulated by cascading the ABCD matrices of the filter

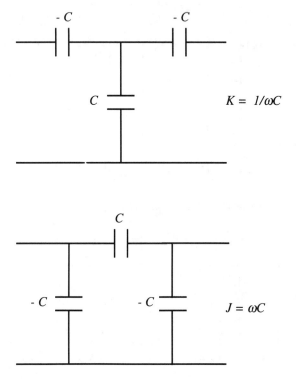

Figure 5.15 Approximation to immittance inverters.

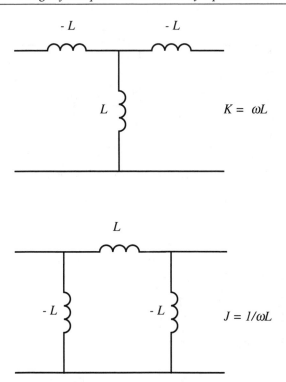

Figure 5.15 (continued).

elements. To illustrate the application of this procedure to the design of a lumped-element lowpass filter, the design of a lowpass filter with the specifications given in Section 5.3 is considered. Figure 5.18 shows the calculated passband return loss (before optimization: dashed line) designed using that procedure. As can be seen, the design specification still is unsatisfactory, and optimization is often required in practice for the accurate design of the filters.

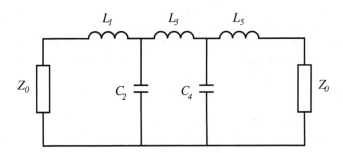

Figure 5.16 Lumped-element lowpass filter with series inductor as first filter element.

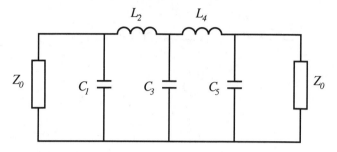

Figure 5.17 Lumped-element lowpass filter with shunt capacitor as first filter element.

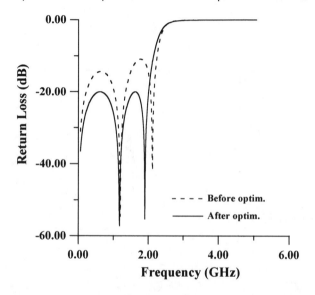

Figure 5.18 Calculated return loss before (dashed line) and after (solid line) optimization of the lowpass lumped-element filter.

5.2 Numerical Implementation of Equal-Ripple Optimization

To apply the equal-ripple optimization technique described in [10,11] to the design of lumped-element lowpass filters, it is necessary, for a given set of filter element values, to be able to calculate the insertion loss on a sample of frequency points within the specified passband. For a lumped-element lowpass filter (see Figure 5.5), the insertion loss and the return loss can be expressed in terms of an ABCD matrix. The matrix representation of the whole filter is

$$
\begin{bmatrix} A & jB \\ jC & D \end{bmatrix} = \begin{bmatrix} A_{L1} & jB_{L1} \\ jC_{L1} & D_{L1} \end{bmatrix} \cdot \begin{bmatrix} A_{C2} & jB_{C2} \\ jC_{C2} & D_{C2} \end{bmatrix}
$$
$$
\cdot \begin{bmatrix} A_{L3} & jB_{L3} \\ jC_{L3} & D_{L3} \end{bmatrix} \cdot \begin{bmatrix} A_{C4} & jB_{C4} \\ jC_{C4} & D_{C4} \end{bmatrix} \cdots \begin{bmatrix} A_{Ln} & jB_{Ln} \\ jC_{Ln} & D_{Ln} \end{bmatrix} \quad (5.33)
$$

in which A_{Li}, B_{Li}, C_{Li}, and D_{Li} are elements of the ABCD matrix of the series inductor as filter element defined by

$$\begin{pmatrix} 1 & \omega L_i \\ 0 & 1 \end{pmatrix} \tag{5.34}$$

and A_{Ci}, B_{Ci}, C_{Ci}, and D_{Ci} are elements of the ABCD matrix of the shunt capacitor as filter element defined by

$$\begin{pmatrix} 1 & 0 \\ \omega C_i & 1 \end{pmatrix} \tag{5.35}$$

The overall filter response—insertion loss (L_I) and return loss (L_R)—can be expressed in terms of elements of the total ABCD matrix of the filter at each frequency (by directly combining the ABCD matrices of the individual filter elements) as

$$L_I = 20 \, \log_{10}\left(\frac{A + B + C + D}{2}\right) \tag{5.36}$$

$$L_R = 20 \, \log_{10}\left(\frac{A + B + C + D}{A + B - C - D}\right) \tag{5.37}$$

A good approximate design of a lumped-element lowpass filter can be obtained by the procedure described in Section 5.1. It is therefore adopted in this chapter as a means of generating a starting point for the optimization.

5.3 Numerical Results

To illustrate our approach, a fifth-order lumped-element lowpass filter has been designed, with the following specifications:

- Cutoff frequency: 2 GHz;
- Passband return loss: 20 dB;
- Insertion loss: 50 dB at 4 GHz;
- Source and load impedances: 50Ω.

The filter can be described by three parameters: inductors (L1, L3) and capacitor (C2), as marked in Figure 5.16, or capacitors (C1, C3) and (L2), as marked in Figure 5.17. We used equal-ripple optimization with L1, L3, and C2 as variables for the filter shown in Figure 5.16, and C1, C3, and L2 for the filter shown in Figure 5.17.

The optimization variables before and after optimization are listed in Tables 5.1 and 5.2. The dashed line in Figure 5.18 shows the calculated passband return loss of filter using the approximate method. This approximate design was used as a starting point for equal-ripple optimization. The passband return loss calculated using the filter elements obtained on convergence are indicated by the solid line in Figure 5.18.

Table 5.1
Lumped-Element Lowpass Filter With Series Inductor as First Filter Element

Parameters	Before Optimization	After Optimization
L_1 (nH):	5.000	3.900
C_2 (pF):	2.000	2.200
L_3 (nH):	8.000	7.200
C_4 (pF):	2.000	2.200
L_5 (nH):	5.000	3.900

Table 5.2
Lumped-Element Lowpass Filter With Shunt Capacitor as First Filter Element

Parameters	Before Optimization	After Optimization
C_1 (pF):	2.000	1.600
L_2 (nH):	5.000	5.500
C_3 (pF):	3.000	2.900
L_4 (nH):	5.000	5.500
C_5 (pF):	2.000	1.600

References

[1] Levy, R., "Design Considerations for Lumped-Element Microwave Filters," *Microwave Journal.*, February 1988, pp. 183–192.

[2] Geffe, P, R., "The Design of Single-Layer Solenoids for RF Filters," *Microwave Journal*, December 1996, pp. 70–76.

[3] Swanson D. G., "Thin-Film Lumped Element Microwave Filters," *IEEE MTT-S Int. Microwave Symp. Dig.*, 1989, pp. 671–674.

[4]   Swanson D. G., R. Porse, and B. J. L. Nilsson, "A 10 GHz Thin-Film Lumped Element High Temperature Superconductor Filter," *IEEE MTT-S Int. Microwave Symp. Dig.*, 1989, pp. 671–674.

[5]   Zhang, D., et al., "Narrowband Lumped-Element Microstrip Filters Using Capacitively-Loaded Inductors," *IEEE Trans. Microwave Theory & Tech.*, Vol. MTT-43, No. 12, December 1995, pp. 3030–3036.

[6]   Matthaei, G. L., S. M. Rohlfing, and R. J. Forse, "Design of HTS, Lumped-Element, Manifold-Type Microwave Multiplexers," *IEEE Trans. Microwave Theory & Tech.*, Vol. MTT-43, No. 7, July 1996, pp. 1313–1321.

[7]   Chi, C-Y., and G. Rebeiz, "Planar Microwave and Millimeter-Wave Lumped Elements and Coupled-Line Filters Using Micro-Machining Techniques," *IEEE Trans. Microwave Theory & Tech.*, Vol. MTT-43, April 1995, pp. 730–738.

[8]   Cohn, S. B., "Direct-Coupled-Resonator Filters," *Proc. IRE*, Vol. 45, February 1957, pp. 187–196.

[9]   Matthaei, G., L. Young, and E. M. T. Jones, *Microwave Filters, Impedance-Matching Networks and Coupling Structures*, Dedham, MA: Artech House, 1980.

[10] Postoyalko, V., and D. Budimir, "Design of Waveguide E-Plane Filters With All-Metal Inserts by Equal-Ripple Optimization," *IEEE Trans. Microwave Theory & Tech.*, Vol. MTT-42, February 1994, pp. 217–222.

[11] *DBFILTER Reference Manual*, Tesla Communications Ltd., London, England.

Selected Bibliography

Bahl, I. J., and P. Bhartia, *Microwave Solid State Circuit Design*, New York: Wiley, 1988.

Bandler, J. W., and S. H. Chen, "Circuit Optimization: The State of the Art," *IEEE Trans. Microwave Theory & Tech.*, Vol. MTT-36, 1988, pp. 424–443.

Chang, K., ed., *Handbook of Microwave and Optical Components, Vol. 1*, New York: Wiley, 1989.

Gaiewski, W. R., L. P. Dunleavy, and L. A. Geis, "Hybrid Inductor Modeling for Successful Filter Design," *IEEE Trans. Microwave Theory & Tech.*, Vol. MTT-42, No. 7, July 1994, pp. 1426–1429.

Gupta, K. C., R. Gary, and R. Chadha, *Computer-Aided Design of Microwave Circuits*, Dedham, MA: Artech House, 1981.

Hasler, M., and J. Neiryuck, *Electrical Filters*, Dedham, MA: Artech House, 1986.

Helszajn, J., *Microwave Planar Passive Circuits and Filters*, Chichester, England: Wiley, 1994.

Puglia, K., and M. Goldfarb, "Designing Lumped Element Bandpass Filters Using Coupled Resonators," *Microwave J.*, September 1989, pp. 197–201.

6

Design of E-Plane Filters by Optimization

Metallic structures placed in the E-plane of a waveguide along the waveguide axis are widely employed in different microwave devices, in particular, frequency selective units. The simplest such structures are waveguide E-plane metal insert bandpass filters (Figure 6.1), with half-wave resonators coupled with one another by means of longitudinally oriented, inductive diaphragms with the rectangular cross-sections (metal septa). These filters offer the potential of realizing low-cost, mass-producible, low-dissipation-loss mm-wave filters [1,2]. E-plane filters can be represented by a normalized equivalent circuit (Figure 6.2), assuming TE_{10} propagation only and neglecting higher order mode coupling between adjacent E-plane septa. Normalization is with respect to the guide impedance. The normalized reactances x_{si}, x_{pi} corresponding to the ith septum ($i = 1, 2, \ldots, n + 1$) are functions of the length of the septum (d_i). Usually the design of these filters is based on half-wave prototypes, with the septa being related to impedance inverters.

The symmetrical T circuit, corresponding to a septum, symmetrically embedded in a length of guide operates as an impedance inverter (Figure 6.3).

When the electrical length ϕ is chosen so that

$$\phi = \_\tan^{-1}(2x_b + x_a) - \tan^{-1}(x_a) \tag{6.1}$$

the normalized ABCD matrix takes the form

$$\begin{bmatrix} 0 & jK \\ j/K & 0 \end{bmatrix} \tag{6.2}$$

125

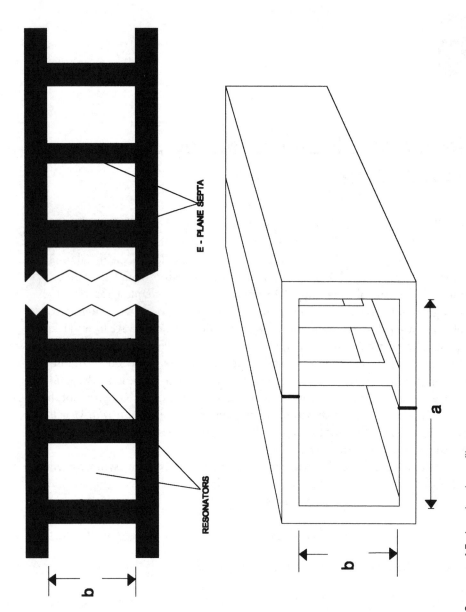

Figure 6.1 Structure of E-plane bandpass filters.

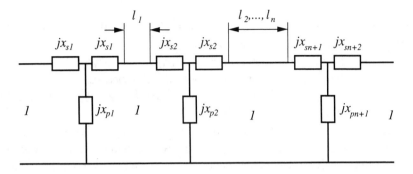

Figure 6.2 Normalized dominant mode equivalent circuit of E-plane filter.

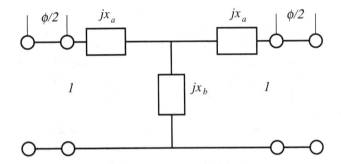

Figure 6.3 *T* circuit embedded in length of guide.

where

$$\tan(2 \tan^{-1} K) = \frac{2x_b}{1 + 2x_b x_a + x_a^2} \tag{6.3}$$

In the case of an E-plane septum, K, defined by (6.3), has a nonlinear frequency dependence and is not constant, although (6.2) has the form of the normalized ABCD matrix of an impedance inverter. Applying (6.1) and (6.3) at the "center" frequency of the specified passband to calculate the l_i and d_i, which correspond to the impedance inverters given by Rhodes' synthesis equations for optimum Chebyshev distributed half-wave prototypes [3], is a common method [1,4–6] in the design of E-plane filters. This approximate treatment of the frequency dependence of (6.1) and (6.3) can result in a designed passband that differs considerably from that which is specified. Therefore, to satisfy the design specification [4], optimization is required to tune the filter dimensions. The accurate calculation of the electrical parameters of E-plane septa requires large computing resources, which makes the optimization of E-plane filters prohibitive. Therefore, in this chapter, a new CAD algorithm is proposed that employs the equal-ripple optimization method described in Chapter 4 for the accurate design of E-plane metal insert bandpass filters. This method is more suited to the problem than standard optimization because it requires less frequency sampling. The elements of the equivalent circuit are calculated using the mode-matching method described in Chapter 3.

Section 6.1 presents the procedure developed by Rhodes for the design of direct-coupled cavity filter. The way in which that procedure attempts to incorporate a linear frequency dependence of the cavity couplings also is discussed.

Lim et al. [1] have attempted to include in an implicit way the actual nonlinear frequency dependence of E-plane septa by applying a passband correction scheme in conjunction with the Rhodes synthesis procedure. That scheme is described in Section 6.2. Section 6.3 describes the application of the equal-ripple optimization method (discussed in Chapter 4) to the design of E-plane filters. The passband correction scheme described in Section 6.2 is used to generate a starting point for the optimization.

A procedure for doing this, requiring only real scalar arithmetic, and numerical aspects of equal-ripple optimization are discussed in Section 6.3. Section 6.5 considers a design example and presents experimental results to confirm the accuracy of the proposed optimization-based design procedure.

All numerical results presented in this work have been obtained using EPFILTER, which is a field theory-based CAD program especially developed for this kind of filter. The software [7] calculates the overall filter response in terms of the even and odd input impedances of the filter structure at each frequency and optimizes the filter parameters. The CAD synthesis software EPFILTER was implemented on SUN workstations (SPARC 10) and IBM PCs. The validity of the new method was confirmed by computer simulations and experimental measurements of the filters designed by this method.

6.1 An Approximate Synthesis-Based Procedure for the Design of Direct-Coupled Cavity Filters

The design procedure outlined in this section is based on a formulation proposed by Rhodes for an optimum Chebyshev distributed stepped impedance lowpass prototype [3]. It is a distributed filter (Figure 6.4a) consisting of a cascade of n line elements (unit elements); each element corresponds to a resonator in the conventional filter design. The elements, having characteristic impedances Z_r ($r = 1, 2, \ldots, n$) are assumed to have an equal length of $\lambda_{g0}/2$, where λ_{g0} is the guide wavelength of the line at the center frequency. The electrical

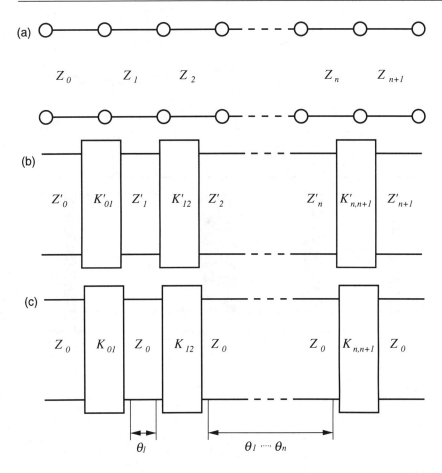

Figure 6.4 (a) Cascade of unit elements; (b) cascade of unit elements and inverters; (c) final network after impedance scaling; (d) E-plane bandpass filter with metal insert; and (e) equivalent circuit using *T*-equivalent for metal septum.

response of this transmission line structure depends on the impedances of the unit elements. For electromagnetic waves propagating along the line, the impedance differences between the unit elements yield reflected waves that, after appropriate arrangement, cancel each other at the desired frequencies. However, if a uniform waveguide is used to implement the circuit, all the unit elements are of the same impedance; the necessary wave reflections must be produced by inserting some sort of discontinuities between the unit elements. One such example is where impedance inverters are inserted as the discontinuities; the unit elements are assumed to have unit impedance as shown in Figure 6.4b. For a cascade of unit elements (see Figure 6.4a), an optimum equal ripple bandpass response occurs around $\theta = \pi$ when

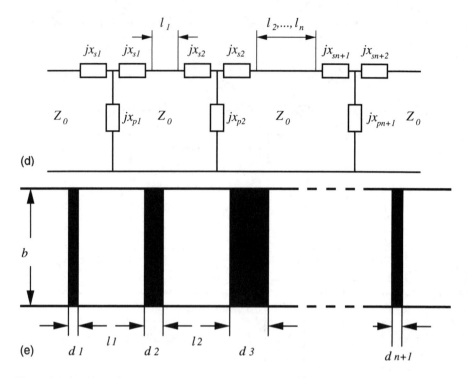

Figure 6.4 (continued).

$$|S_{12}|^2 = \frac{1}{1 + \epsilon^2 \, T_n^2(a \sin \, \theta)} \tag{6.4}$$

where

$$\theta = \frac{\lambda_{g0}}{\lambda_g} \tag{6.5}$$

and

$$T_n(x) = \cosh[n \, \text{Arcosh}(x)] \quad \text{for } |x| > 1 \tag{6.6a}$$

$$T_n(x) = \cos[n \, \arccos(x)] \quad \text{for } 0 < x \le 1 \tag{6.6b}$$

with

$$x = x \sin \, \theta \tag{6.7}$$

is the nth degree Chebyshev polynomial of the first kind. α defines the passband bandwidth (scaling parameter), ϵ defines the passband ripple level, λ_{g0} is the guide wavelength at the center frequency, and λ_g is the guide wavelength. In this discussion, the discontinuities are assumed to be frequency independent, which is hardly true in practice. The frequency-dependent behavior of the discontinuity has significant effects on the filter performance. In an earlier paper, Levy [8] studied in detail the reactance coupled filters and concluded that the response in (6.4) should be modified as

$$|S_{12}|^2 = \cfrac{1}{1 + \epsilon^2 \, T_n^2\left[\dfrac{\alpha \, \lambda_g}{\lambda_{g0}} \sin\left(\dfrac{\pi \, \lambda_{g0}}{\lambda_g}\right)\right]} \tag{6.8}$$

to take into account the linear frequency dependence of the discontinuities.

To design a filter, the designer normally is given the two passband edge frequencies yielding λ_{gL} and λ_{gH}, passband return loss (L_R), stopband attenuation (L_I), the waveguide housing dimensions (a,b), and the metal-septum thickness (t). The design procedure is as follows:

1. From λ_{gL} and λ_{gH}, determine the parameters α and λ_{g0} from the following nonlinear equations, which readily can be solved numerically [3]:

$$\frac{\alpha \, \lambda_{gL}}{\lambda_{g0}} \sin\left(\frac{\pi \, \lambda_{g0}}{\lambda_{gL}}\right) = 1$$

$$\frac{\alpha \, \lambda_{gH}}{\lambda_{g0}} \sin\left(\frac{\pi \, \lambda_{gH}}{\lambda_{gH}}\right) = -1 \tag{6.9}$$

where λ_{g2} and λ_{g1} are the guide wavelengths at the upper and lower bandedge frequencies, respectively. Equation (6.9) comes from the property of the Chebyshev polynomial of the first kind, which oscillates between +1 and −1. Adding the two parts of (6.9) gives

$$\frac{\alpha \, \lambda_{gL}}{\Delta(f_1)} \sin\left(\frac{\pi \, \lambda_{g0}}{\lambda_{gL}}\right) + \frac{\alpha \, \lambda_{gH}}{\Delta(f_2)} \sin\left(\frac{\pi \, \lambda_{g0}}{\lambda_{gH}}\right) = 0 \tag{6.10}$$

If $\lambda_{g1} \approx \lambda_{g2} \approx \lambda_{g0}$, then we can approximate (6.10) as

$$\frac{\alpha \, \lambda_{gL}}{\Delta(f_1)} \, \pi\left(1 - \frac{\lambda_{g0}}{\lambda_{gL}}\right) + \frac{\alpha \, \lambda_{gH}}{\Delta(f_2)} \, \pi\left(1 - \frac{\lambda_{g0}}{\lambda_{gH}}\right) = 0 \tag{6.11}$$

giving

$$\lambda_{g0} = \frac{\lambda_{gL} + \lambda_{gH}}{2} \qquad (6.12)$$

However, for broad bandwidths, this will not be sufficiently accurate. A better solution is obtained by applying the Newton-Raphson technique, as follows. Let

$$F^*(\lambda_{g0}) = \lambda_{gL} \sin\left(\frac{\pi\,\lambda_{g0}}{\lambda_{gL}}\right) + \lambda_{gH} \sin\left(\frac{\pi\,\lambda_{g0}}{\lambda_{gH}}\right) \qquad (6.13)$$

Then differentiating w.r.t. λ_{g0} gives

$$F^*(\lambda_{g0}) = \lambda_{gL} \sin\left(\frac{\pi\,\lambda_{g0}}{\lambda_{gL}}\right) + \lambda_{gH} \sin\left(\frac{\pi\,\lambda_{g0}}{\lambda_{gH}}\right) \qquad (6.14)$$

and the modified value of λ_{g0} will be

$$\lambda_{g0}^{new} = \lambda_{gL} - \frac{F^*(\lambda_{g0})}{F'(\lambda_{g0})} \qquad (6.15)$$

2. Determine the passband ripple level from the minimum passband return loss, which is defined as:

$$L_R = 10 \log_{10}\left(1 + \frac{1}{\epsilon^2}\right) \qquad (6.16)$$

3. Determine the number of resonators n from (6.17). That is accomplished by finding the minimum value of n for which the most severe constraint on the stopband insertion loss level (L_I) satisfies

$$L_I = 10 \log_{10}\left\{1 + \epsilon^2\, T_n^2\left[\frac{\alpha\,\lambda_g}{\lambda_{g0}} \sin\left(\frac{\pi\,\lambda_{g0}}{\lambda_g}\right)\right]\right\} \qquad (6.17)$$

at the designated stopband frequency f_s. In (6.17), λ_g is the guide wavelength at f_s.

4. For a cascade of unit elements with the transfer characteristic given by (6.4), we can synthesize the network to obtain the characteristic

impedances of the unit elements. However, for up to moderate band-widths, we can use explicit formulas for the element values [9]. Modifying the network by introducing impedance inverters of the characteristic impedances $K_{r,r+1}$, as shown in Figure 6.4b, the explicit formulas for the impedance values are

$$Z'_r = \frac{2\alpha}{y} \sin\left[\frac{(2r-1)\,\pi}{2n}\right] - \frac{1}{4\alpha y}\left[\frac{y^2 + \sin^2\left(\frac{r\pi}{n}\right)}{\sin\frac{(2r+1)\pi}{2n}}\right]$$

$$-\frac{1}{4\alpha y}\left[\frac{y^2 + \sin^2\frac{(r-1)\pi}{n}}{\sin\frac{(2r-3)\pi}{2n}}\right] \quad r = 1, 2, \ldots, n \quad (6.18)$$

where

$$K'_{r,r+1} = \frac{\sqrt{y^2 + \sin^2\left(\frac{n\pi}{n}\right)}}{y} \quad r = 0, 1, \ldots, n \quad (6.19)$$

with

$$y = \sinh\left(\frac{1}{n}\sinh^{-1}\frac{1}{\epsilon}\right) \quad (6.20)$$

5. In the rectangular waveguide realization, because the waveguide is uniform, we must scale the internal impedance level, as shown in Figure 6.4c, where

$$K_{r,r+1} = \frac{K'_{r,r+1}}{\sqrt{Z'_r Z'_{r+1}}} \quad r = 0, 1, \ldots, n \quad (6.21)$$

with

$$Z'_n = Z'_{n+1} = 1 \quad (6.22)$$

6. Determine the *i*th septum length, d_i (Figure 6.4d) by solving (6.3), so that the required impedance inverter is realized. The normalized

reactances x_a and x_b (Figure 6.4e) are a function of the septum width. Because these functions are not available explicitly, we must implement a root-seeking routine to find the value of width that is provided by the required impedance value K and the angle f for each impedance inverter.

7. Finally, the length of the ith resonator (l_i) (see Figure 6.4e) formed by ith and ($i + 1$)th septa is given by

$$l_i = \frac{\lambda_{g0}}{2\pi}\left[\pi - \frac{1}{2}(\phi_i + \phi_{i+1}) \right] \qquad (6.23)$$

where ϕ_i is given by (6.1), and the electrical distance, π, corresponds to the physical distance, $\lambda_{g0}/2$.

The range of validity for the E-plane filter design is limited because of the inaccurate approximation of the E-plane septa made in the derivation of the design procedure. This problem can be reduced by using the method proposed by Lim et al. [1]. That method is described in Section 6.2.

6.2 A CAD Algorithm Using the Passband Correction Method

This section describes the passband correction scheme proposed by Lim et al. [4] to incorporate the actual nonlinear frequency dependence of the E-plane septa into the Rhodes synthesis procedure.

This scheme compensates the linear frequency dependence of E-plane septa to reduce the passband deviation problem associated with the procedure described in Section 6.1. Equation (6.8) predicts the response of E-plane bandpass filters more accurately if a modification of linear frequency dependence by correction factors is introduced. Those correction factors implicitly include the actual frequency dependence of the E-plane septa. In E-plane filters, the correction factor cannot be expressed in an analytical form, because such a quantity is a complicated function of both frequency and E-plane septa widths. To define the passband correction factor, we used a numerical approach instead of an analytical approach. Starting with a filter predesigned by the method described in Section 6.1, we can express the passband correction factor as a function of frequency only. In this algorithm, only two passband correction factors are required and are calculated from the actual insertion losses (L_I) of the predesigned filter at the specified bandedge frequencies (f_L and f_H). When the passband ripple, the lower and upper bandedge frequencies, f_L and f_H, and the number of resonators are given, the insertion loss characteristic of a waveguide bandpass filter with a Chebyshev characteristic [1] is given by

$$L_I = 10 \log_{10}\left[1 + \epsilon^2 \, T_n^2\!\left(\frac{\alpha}{m}\right)\right] \tag{6.24}$$

where $m = \lambda_{g0}/\lambda_g$ is the frequency dependence of the K-inverters. The factor α and the guide-wavelength λ_{g0} at the center frequency are determined by the procedure given in Section 6.1.

The actual insertion losses L_{IL} and L_{IH} (in decibels) at f_L and f_H are found by calculating the filter response by a field theoretical method, specifically the mode-matching method developed in this work. In practice, the difference between the septa width of a predesigned filter and those of a final filter is found to be so small that the frequency dependence is assumed to be independent of the septa width variations. Thus, starting with a filter predesigned by the method given in Section 6.1, we can represent the passband correction factor as a function of frequency only as described next.

It is assumed that the actual frequency dependence of the E-plane discontinuity is $M = m\Delta(f)$, where $\Delta(f)$ is defined as the passband correction factor that we try to find. Then the actual insertion losses (L_I) of the predesigned filter are represented by

$$L_{I \, actual} = 10 \log_{10}\left\{1 + \epsilon^2 \, T_n^2\!\left[\frac{\alpha}{M} \sin\!\left(\frac{\pi \, \lambda_{g0}}{\lambda_g}\right)\right]\right\}$$

$$= 10 \log_{10}\left\{1 + \epsilon^2 \, T_n^2\!\left[\frac{\alpha \, \lambda_g}{\Delta(f)\lambda_{g0}} \sin\!\left(\frac{\pi \, \lambda_{g0}}{\lambda_g}\right)\right]\right\} \tag{6.25}$$

where λ_{g0} and α have the same values as in (6.24). At the specified bandedge frequencies, (6.25) must satisfy the following condition:

$$10 \log_{10}\left[1 + \epsilon^2 \, T_n^2\!\left(\frac{1}{\Delta(f_i)}\right)\right] = L_{I \, actual} \quad i = L, H \tag{6.26}$$

From (6.26), we can calculate the passband correction factors, $\Delta(f_L)$ and $\Delta(f_H)$, at the specified bandedge frequencies as

$$\Delta(f_i) = \cfrac{1}{\cosh\!\left(\cfrac{1}{n} \ln(cst_i + \sqrt{cst_i^2 - 1})\right)} \quad i = L, H \tag{6.27}$$

where

$$cst_i = \frac{10^{0.1L_i} - 1}{\epsilon^2} \qquad i = L, H \tag{6.28}$$

Because our final goal is that the actual insertion loss-characteristics $L_{I\ actual}$ of the final E-plane filter have the same passband characteristics as the L_I of (6.24), which is the given specification, λ_{g0} and α in (6.25) must be recalculated from (6.34) and (6.35), so that (6.25) satisfies the specified passband ripple, $10 \log_{10}(1 + \epsilon^2)$, at both f_L and f_H.

$$\frac{\alpha\,\lambda_{gL}}{\Delta(f_L)} \sin\left(\frac{\pi\,\lambda_{g0}}{\lambda_{gL}}\right) + \frac{\alpha\,\lambda_{gH}}{\Delta(f_H)} \sin\left(\frac{\pi\,\lambda_{g0}}{\lambda_{gH}}\right) = 0 \tag{6.29}$$

If $\lambda_{gL} \approx \lambda_{gH} \approx \lambda_{g0}$, then we can approximate as

$$\frac{\alpha\,\lambda_{gL}}{\Delta(f_L)}\,\pi\left(1 - \frac{\lambda_{g0}}{\lambda_{gL}}\right) + \frac{\alpha\,\lambda_{gH}}{\Delta(f_H)}\,\pi\left(1 - \frac{\lambda_{g0}}{\lambda_{gH}}\right) \tag{6.30}$$

giving

$$\lambda_{g0} = \frac{\lambda_{gL}\,\Delta(f_H) + \lambda_{gH}\,\Delta(f_L)}{\Delta(f_L) + \Delta(f_H)} \tag{6.31}$$

For broad bandwidths, however, that is not sufficiently accurate. A better solution is obtained by applying the Newton-Raphson technique, as follows. Let

$$F^*(\lambda_{g0}) = \frac{\lambda_{gL}}{\Delta(f_L)} \sin\left(\frac{\pi\,\lambda_{g0}}{\lambda_{gL}}\right) + \frac{\lambda_{gH}}{\Delta(f_H)} \sin\left(\frac{\pi\,\lambda_{g0}}{\lambda_{gH}}\right) \tag{6.32}$$

Then differentiating w.r.t. λ_{g0} gives

$$F'(\lambda_{g0}) = \frac{\pi}{\Delta(f_L)} \cos\left(\frac{\pi\,\lambda_{g0}}{\lambda_{gL}}\right) + \frac{\pi}{\Delta(f_H)} \cos\left(\frac{\pi\,\lambda_{g0}}{\lambda_{gH}}\right) \tag{6.33}$$

and the modified values of λ_{g0} and α are

$$\lambda_{g0}^{new} = \lambda_{gL} - \frac{F^*(\lambda_{g0})}{F'(\lambda_{g0})} \tag{6.34}$$

$$\alpha^{new} = \frac{\Delta(f_L)}{\dfrac{\lambda_{gL}}{\lambda_{g0}^{new}} \sin\left(\dfrac{\pi \lambda_{g0}^{new}}{\lambda_{gL}}\right)}$$

$$= -\frac{\Delta(f_H)}{\dfrac{\lambda_{gH}}{\lambda_{g0}^{new}} \sin\left(\dfrac{\pi \lambda_{g0}^{new}}{\lambda_{gH}}\right)} \qquad (6.35)$$

With the new λ_{g0} and α, (6.25) accurately predicts the insertion loss characteristics of the final filter around the passband, which are coincident with the prediction of (6.24).

6.2.1 Design Procedure Using Passband Correction Method

For a given specification, the design procedure of E-plane bandpass filters using the proposed passband correction method is summarized as follows:

1. Predesign the E-plane bandpass filter using the procedures given in Section 6.1.

2. Using a field theoretical analysis method, calculate the actual insertion losses L_{IL} and L_{IH} of the predesigned E-plane bandpass filter at the specified bandedge frequencies f_L and f_H.

3. Calculate the passband correction factors $\Delta(f_L)$ and $\Delta(f_H)$, using (6.27).

4. Calculate the new λ_{g0} and the new α from (6.34) and (6.35), respectively.

5. Finally, complete the design of the E-plane bandpass filter using the procedure given in Section 6.1 with the new λ_{g0} and α.

To illustrate the application of this procedure to the design of E-plane filters, the design of an E-plane bandpass filter with the specifications given in Section 6.6 is considered. Figure 6.5 shows the calculated passband insertion loss of an E-plane bandpass filter designed using the procedure. As can be seen, the design specification still is unsatisfactory, and optimization often is required in practice for the accurate design of these filters.

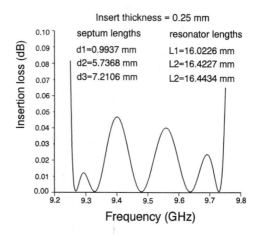

Insert thickness = 0.25 mm

septum lengths	resonator lengths
d1=0.9937 mm	L1=16.0226 mm
d2=5.7368 mm	L2=16.4227 mm
d3=7.2106 mm	L2=16.4434 mm

Figure 6.5 Calculated insertion loss before optimization.

6.3 Numerical Implementation of Equal-Ripple Optimization

6.3.1 Calculation of Insertion Loss

To design E-plane filters for a given set of metal insert dimensions using the equal-ripple optimization method described in Chapter 4, it is necessary to be able to calculate the insertion loss on a sample of frequency points within the specified passband. In the case of longitudinally symmetrical structures, such as conventional E-plane bandpass filters, the insertion loss (L_I) can be expressed in terms of normalized even- and odd-mode impedances as

$$L_I = 20 \log_{10} \frac{(1 + z_e)(1 + z_0)}{z_e - z_0} \qquad (6.36a)$$

where $jz_{e(0)}$ is the normalized input impedance of the two identical one-ports formed by placing a magnetic (electric) wall at the plane of symmetry. By transforming an open (short) circuit placed at the plane of symmetry through the filter sections (resonators and E-plane septa) located to the left of the plane of symmetry, $z_{e(0)}$ can be calculated. Each E-plane septum is itself symmetrical and can be electrically represented by normalized even- and odd-mode impedances

$$z_{ei} = j(x_{si} + 2x_{pi}) \qquad (6.36b)$$

$$z_{0i} = j\, x_{si} \qquad (6.36c)$$

For a normalized reactive impedance jz, an E-plane septum performs the normalized impedance transformation $jz \Rightarrow jz_{in}$, where z_{in} is given by

$$z_{in} = \frac{z(z_{ei} + z_{oi}) + 2\,z_{ei}z_{oi}}{2z + (z_{ei} + z_{oi})} \qquad (6.36d)$$

A resonator section, that is, a length of guide, performs the normalized impedance transformation $jz \Rightarrow jz_{in}$, where z_{in} is given by

$$z_{in} = \frac{z + \tan \beta l}{1 - z \tan \beta l} \qquad (6.36e)$$

where β ($= 2\pi/\lambda_g$) is the propagation constant, and l is the length of the resonator. By applying (6.36d) and (6.36e), it is possible to calculate z_e and z_0 starting at the center of the filter and working outward. That process involves no matrix manipulation and uses only real arithmetic. For the analysis of E-plane septa in a rectangular waveguide, the mode-matching method described in Section 3.1 was used.

Neither accurate numerically fitted closed-form expressions nor accurate design tables for the electrical parameters of E-plane septa in terms of dimensions (length and thickness) and frequency are yet available [5,10]. Therefore, the accurate design of E-plane filters requires the direct calculation of the electrical parameters of E-plane septa and highlights the need for the design of E-plane filters by optimization techniques. Optimization techniques minimize the number of calculations of the electrical parameters of E-plane septa.

6.3.2 Solution of Nonlinear Equations

For the solution of a system of nonlinear equations, the Newton-Raphson method is a rapidly convergent technique if a good initial approximation is available. A good approximate design of an E-plane filter can be obtained by the procedure given in Section 6.2. It is therefore adopted in this work as a means of generating a starting point for the solution of a system of $n + 1$ nonlinear equations in $n + 1$ variables [see (4.13)] by means of the Newton-Raphson method.

The number of times the function is evaluated in the process of finding its root is the usual measure of computational effort. That includes function evaluations required to calculate derivatives numerically. The most complex task in the Newton-Raphson method is evaluating the Jacobian J of $E(x)$. The Jacobian matrix of the $n + 1$ nonlinear functions can be calculated numerically using finite difference as

$$[J_{ij}] = \frac{\partial E_i}{\partial x_j} \cong \frac{E_i(x_j + \delta x_j) - E_i(x_j)}{\delta x_j}$$

$$i = 1, 2, \ldots, n + 1; \, j = 1, 2, \ldots, n + 1 \tag{6.37}$$

The finite difference calculation of the Jacobian matrix for a given set of filter parameters requires the evaluation of

$$E_i(x_1, x_2, \ldots, x_j + \delta x_j, \ldots, x_{n+1})$$

$$i = 1, 2, \ldots, n + 1; \, j = 1, 2, \ldots, n + 1 \tag{6.38}$$

where $x_1, x_2, \ldots, x_{n+1}$ denote the $n + 1$ dimensions required to specify a symmetrical E-plane filter. For odd n:

$$x_i = d_i \quad i = 1, 2, \ldots, \frac{n + 1}{2} \tag{6.39}$$

$$x_i = l_i - \frac{(n + 1)}{2} \quad i = \frac{n + 3}{2}, \frac{n + 5}{2}, \ldots, n + 1 \tag{6.40}$$

For even n:

$$x_i = d_i \quad i = 1, 2, \ldots, \frac{n}{2} + 1 \tag{6.41}$$

$$x_i = l_i - \frac{(n + 1)}{2} \quad i = \frac{n}{2} + 2, \frac{n}{2} + 3, \ldots, n + 1 \tag{6.42}$$

Denoting by x and E the $n + 1$ dimensional vectors with components $x_i(i = 1, 2, \ldots, n + 1)$ and $E_i(i = 1, \ldots, n + 1)$, the Newton-Raphson method has the general form [11]

$$x^k = x^{k-1} - J^{-1}(x^{k-1})E(x^{k-1}) \tag{6.43}$$

where k is the iteration number ($k = 1, 2, \ldots$) and J^{-1} is the inverse of the m-by-m Jacobian matrix evaluated at x^{k-1}. After each iteration, the response and errors are computed again with the new corrected parameters, until the errors are judged to be sufficiently small.

    To minimize the CPU time required to evaluate the Jacobian matrix, the normalized even- and odd-mode impedances corresponding to the septum lengths at the start of a Newton-Raphson iteration should be calculated at the sample frequencies and stored. That avoids unnecessary recalculation of the parameters when the insert dimensions are perturbed in turn to calculate the elements of the Jacobian matrix.

6.4 Numerical and Experimental Results

To verify the accuracy of the developed method, the design of an E-plane bandpass filter with the following specifications was considered:

- Waveguide WG16 (WR90) internal dimension: 22.86 mm by 10.16 mm;
- Midband frequency: 9.5 GHz;
- Passband: 9.25–9.75 GHz;
- Ripple level: 0.05 dB;
- Number of resonators: 5;
- Metal insert thickness: 0.250 mm;
- Passband return loss: 20 dB min;
- Filter characteristic: Chebyshev.

The calculated passband insertion loss of an E-plane filter designed using the approximate method given in Section 6.2 is shown in Figure 6.5. Also included in that figure are dimensions of the E-plane insert. In both the design and the calculations, mode matching with 100 modes was used. As a starting point for equal-ripple optimization, this approximate design was used. The passband insertion loss calculated using the insert dimensions obtained on convergence is given in Figure 6.6. This took four Newton-Raphson iterations. Mode matching with 100 modes was used throughout the optimization.

The calculated insertion loss of the final design over the whole waveguide band is shown in Figure 6.7 (dashed line). This figure (solid line) also shows a plot of (6.8) with $n = 5$, $\epsilon^2 = 10^{.005} - 1$, $\lambda_{g0} = 43.808$ mm and $\alpha = 6.344$. Although the passband to stopband discrimination of the designed filter is similar to that predicted by (6.8), it has inferior stopband attenuation in the upper part of the waveguide band. That limitation of conventional E-plane filters is well known [2,7,10]. Improved stopband attenuation can be achieved by using solutions presented in Chapter 7. To evaluate the effect of higher order mode interaction on this design, the passband insertion loss was recalculated, again using the dimensions given in Figure 6.6. In that recalculation of

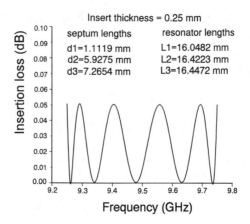

Figure 6.6 Calculated insertion loss after optimization.

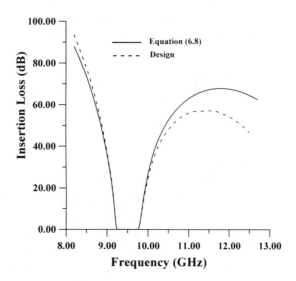

Figure 6.7 Comparison of calculated insertion loss with (6.8).

the passband insertion loss, the TE_{30} coupling between E-plane septa was taken into account. The recalculated insertion loss displayed no noticeable difference from that shown in Figure 6.6.

The measured insertion loss of the fabricated design of a five-resonator E-plane bandpass filter is shown in Figure 6.8. The measurement was made using an HP 8510C vector network analyzer. A full two-port calibration was used. Waveguide standards calling for a short, offset short, sliding load, and through were used during the calibration. The filter design was fabricated using

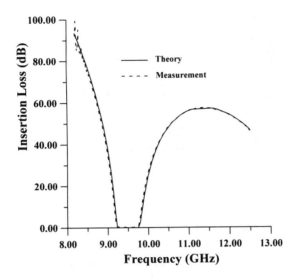

Figure 6.8 Measured (dashed line) and calculated (solid line) insertion loss.

a brass waveguide housing and a copper insert. The metal insert was realized by using spark erosion. The measured insertion loss in the passband was less than 0.45 dB. The discrepancies between the measured insertion loss and theory within the passband are due to ohmic losses in metallic walls and metal insert.

That the response of the fabricated filter is equal ripple can be seen from the measured return loss, which is shown in Figure 6.9. Compared with the theoretical prediction, the passband of the fabricated filter is shifted downward slightly. Considering the inaccuracies involved in the practical realization of E-plane filters [2,7,10], the agreement between theory and measurement is very good. The comparison of the measured and simulated filter responses shows slight disagreement (bandpass is shifted). That disagreement may be due to spark erosion errors.

Figure 6.10 is a photograph of a five-resonator metal insert E-plane bandpass filter 1 structure together with the corresponding waveguide housing at X-band.

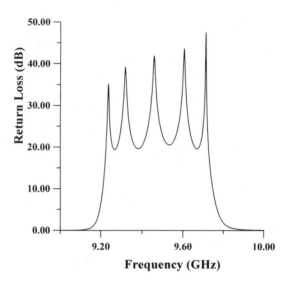

Figure 6.9 Measured return loss.

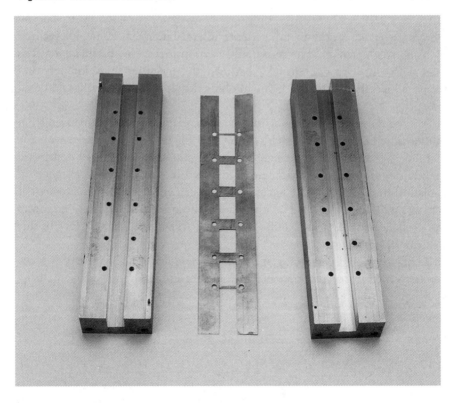

Figure 6.10 Photograph of conventional E-plane bandpass filter.

References

[1] Lim, J. B., C. W. Lee, and T. Itoh, "An Accurate CAD Algorithm for E-Plane Type Bandpass Filters Using a New Passband Correction Method Combined With the Synthesis Procedures," *IEEE MTT-S Int. Microwave Symp. Dig.*, June 1990, pp. 1179–1182.

[2] Arndt, F., "The Status of Rigorous Design of Millimetre Wave Low Insertion Loss Fin-Line and Metalic E-Plane Filters," *J. Instn. Electronics and Telecom. Engrs.*, Vol. 34, No. 2, 1988, pp. 107–119.

[3] Rhodes, J. D., "Microwave Circuit Realizations," in *Microwave Solid State Devices and Applications*, D. V. Morgan and M. J. Howes, eds., England: Peregrinus, 1980, pp. 49–57.

[4] Shih, Y. C., "Design of Waveguide E-Plane Filters With All Metal Inserts," *IEEE Trans. Microwave Theory & Tech.*, Vol. MTT-32, July 1984, pp. 695–704.

[5] Hong, J. S., "Design of E-Plane Filters Made Easy," *IEE Proc.*, Vol. 136, Pt. H, 1989, pp. 215–218.

[6] Bui, L. Q., D. Ball, and T. Itoh, "Broad-Band Millimeter-Wave E-Plane Bandpass Filters," *IEEE Trans. Microwave Theory & Tech.*, Vol. MTT-32, December 1984, pp. 1655–1658.

[7] DBFILTER *Reference Manual*, Tesla Communications Ltd., London, England.

[8] Levy, R., "Theory of Direct Coupled Cavity Filters," *IEEE Trans. Microwave Theory & Tech.*, Vol. MTT-15, June 1967, pp. 340–348.

[9] Rhodes, J. D., *Theory of Electric Filters*, New York: Wiley, 1976.

[10] Gololobov, V. P., and M. Yu. Omel'yanenko, "Bandpass Filters Based on Planar Metal-Dielectric Structures in the E-Plane of a Rectangular Waveguide (A Review)," *Radioelectronics and Communications Systems*, Vol. 30, No. 1, 1987, pp. 1–15.

[11] Ortega, J. M., and W. C. Rheinboldt, *Iterative Solution of Nonlinear Equations in Several Variables*, New York: Academic Press, 1970.

Selected Bibliography

Cohn, S. B., "Direct Coupled Resonator Filters," *Proc. IRE*, Vol. 45, February 1957, pp. 187–196.

Konishi, Y., and K. Uenakada, "The Design of Bandpass Filter With Inductive Strip-Planar Circuit Mounted in Waveguide," *IEEE Trans. Microwave Theory & Tech.*, Vol. MTT-22, October 1974, pp. 869–873.

Levy, R., "Tables of Element Values for the Distributed Low-Pass Prototype Filter," *IEEE Trans. Microwave Theory & Tech.*, Vol. MTT-13, September 1965, pp. 514–536.

Matthaei, G., L. Young, and E. M. T. Jones, *Microwave Filters, Impedance—Matching Networks and Coupling Structures*, Dedham, MA: Artech House, 1980.

Politi, M., et al., "An Equivalent Circuit for the Design of E-Plane Metal-Insert Filters in Millimeter-Wave Applications," *Proc. 20th European Microwave Conf.*, Budapest, Hungary, 1990, pp. 1257–1262.

Rhodes, J. D., "Design Formulas for Stepped Impedance Distributed and Digital Wave Maximally Flat and Chebyshev Low-Pass Prototype Filters," *IEEE Trans. Circuits Syst.*, Vol. CAS-22, November 1975, pp. 866–874.

7

Design of Ridged Waveguide Filters by Optimization

All-metal inserts placed in the E-plane of a rectangular waveguide along the waveguide axis offer the potential of realizing low-cost, mass-producible, and low-dissipation-loss mm-wave filters [1-3]. However, despite their favorable characteristics, the attenuation in the second stopband (i.e., where the resonators are about one wavelength long) often may be too low and too narrow for many applications, such as for diplexers, in which frequency selectivity and high stopband attenuation are considered to be important filtering properties. That is due to the effect that is characteristic for the common single insert design: beyond the cutoff frequency of the fundamental mode within the septum section, which is determined by the distance between the septa and the waveguide sidewalls, the power is increasingly transported directly by propagating waves [4], causing degradation of the selective properties of resonators based on two septa and the waveguide between them.

In recent years, much effort has been devoted to the study of E-plane bandpass filters with improved stopband performance. Several solutions have been proposed:

- Reduce the distance between the metal insert and the waveguide sidewalls by a thick metal insert [5] or use several metal inserts rather than a single insert [5–7]. The first approach achieves good stopband performance, as is indicated in [5], but it has the disadvantage of high passband insertion loss. The latter approach also achieves good stopband performance, but it requires greater effort in the mounting and adjustment of the several metal inserts in the waveguide.

147

- Use narrower or wider waveguide for the filter section, depending on the position of the filter passband within the waveguide single-mode bandwidth [4]. The reduction of the waveguide housing width decreases the distance between the metal insert and the waveguide sidewalls so that the propagation of modes along the coupling subsections is suppressed up to higher frequencies. That results in an improved stopband attenuation when the passband is not the high end of the waveguide band. Increasing the waveguide housing width reduces the guide wavelength of the filter resonators, resulting in an improved stopband performance when the passband is close to the cutoff frequency of the standard width guide.

- Use resonators of different cutoff frequency [8,9]. That is achieved by employing sections of rectangular waveguide in which all the waveguide sections are resonant at the same fundamental frequency. However, due to different guide wavelengths in the different sections, the sections are not all simultaneously resonant at any higher frequency, resulting in an improved stopband attenuation.

This chapter investigates a new solution for improvement in the second stopband. Section 7.1 presents the proposed filter configuration. Section 7.2 describes the circuit representation and design procedure for the design of E-plane bandpass filters with improved stopband performance (ridged waveguide filters). Section 7.3 discusses the efficient computer implementation of equal-ripple optimization, while Section 7.4 considers design examples to confirm the improvement and presents experimental results to confirm the accuracy of the design procedure.

7.1 Proposed Filter Configuration

Based on the existing idea to use different waveguide resonators with different characteristic impedances and different cutoff frequencies [4], improved stopband performance can be met by the E-plane filter configurations shown in Figure 7.1.

That improved performed is due to the nonlinear relation between guide wavelength and frequency, which can be influenced favorably by a suitable reduction of the cutoff frequency of the fundamental mode within the filter resonators. The proposed filter configuration is constructed of direct-coupled ridged waveguide sections that have, in general, identical cutoff frequencies and characteristic impedances, and reactive elements (metal septa) arranged in

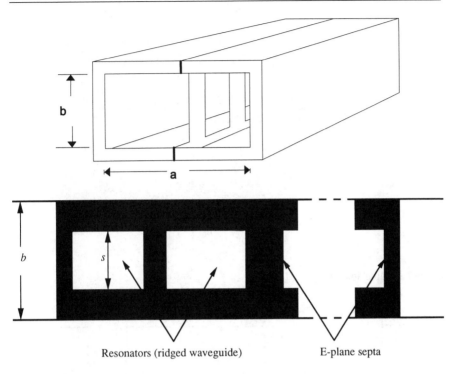

Figure 7.1 Configuration of ridged waveguide filter structure.

such a manner that each section is resonant at the same fundamental frequency. The main features of the new structure are the use of conventional rectangular waveguide housing and the use of a metal insert that, when mounted, introduces ridges in the resonators. The improvement in the upper stopband associated with the superior electrical performance of ridged waveguide, such as cutoff frequency reduction, provides a convenient way to realize E-plane bandpass filters with improved stopband performance. The structure is simple and compatible with the E-plane manufacturing process.

7.2 Circuit Representation and Design Procedure

7.2.1 Circuit Representation

The proposed filter structure in Figure 7.1 can be represented as shown in Figure 7.2 by use of the asymmetrical impedance inverter illustrated in Figure 7.3. The design of these filters usually is based on the design procedure described in Section 7.2.2, with the septa being related to impedance inverters. Figure 7.3 shows a two-port defined by its ABCD matrix. We assume that it

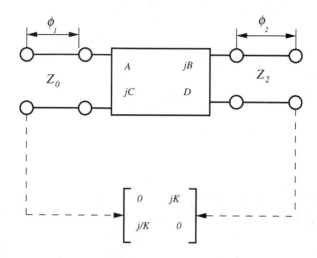

Figure 7.2 Impedance inverter.

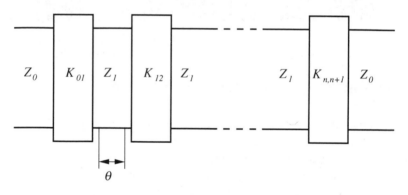

Figure 7.3 Distributed half-wave prototype using impedance inverters.

is connected to lines of characteristic impedances Z_1 and Z_2 at its two-ports. It is always possible to find reference planes P_1 and P_2 at electrical distances ϕ_1 and ϕ_2 from the respective ports to convert the asymmetric two-port (metal septum in rectangular waveguide between two different ridged waveguides) to the symmetric impedance inverter K. The equations giving the reference plane locations are in terms of the ABCD matrix parameters, leading to the results that follow.

The normalized element value of the impedance inverter (see Figure 7.3) can be derived directly from the normalized ABCD matrix of the septum. Normalization is with respect to the guide impedance of the rectangular waveguide. It is given by

$$K = \sqrt{Z_1 Z_2} \left(\sqrt{L} - \sqrt{L-1} \right) \tag{7.1}$$

with

$$L = 1 + \frac{1}{4}[(a-d)^2 + (b-c)^2] \tag{7.2}$$

where a, b, c, and d are normalized elements of the ABCD matrix given as

$$a = A\sqrt{\frac{Z_2}{Z_1}} \quad b = \frac{B}{\sqrt{Z_1 Z_2}} \tag{7.3}$$

$$c = C\sqrt{Z_1 Z_2} \quad d = D\sqrt{\frac{Z_1}{Z_2}} \tag{7.4}$$

and Z_1 and Z_2 are normalized guide impedances of the ridge waveguide resonator sections, which are frequency dependent.

The reference plane locations are given by the equations

$$\tan(2\phi_1) = \frac{2(bd - ac)}{(a^2 - d^2) + (b^2 - c^2)} \tag{7.5}$$

$$\tan(2\phi_2) = \frac{2(ab - cd)}{(d^2 - a^2) + (b^2 - c^2)} \tag{7.6}$$

In practice, electrical distances (ϕ_1 and ϕ_2) can be realized as negative values in the adjacent positive line length, which therefore becomes shortened in the final network. Mathematical details of the derivations (K, ϕ_1, and ϕ_2) are given in the appendix to this chapter. Although an impedance inverter has the form of a normalized ABCD matrix, in the case of an E-plane septum, K, as defined by (7.1), is not constant and has a nonlinear frequency dependence.

7.2.2 Design Procedure

A common approach to the design of the conventional E-plane bandpass filters (as described in [10–12]), can be used with minor modifications for filter structures with different impedances and cutoff frequencies such as E-plane bandpass filters with improved stopband performance. Here, only the most important steps in the design procedure, which should include the concept of impedance inverters and impedance scaling of the impedance levels of the prototype filter, are presented. The design procedure for the E-plane bandpass filters with improved stopband performance is to apply (7.1), (7.5), and (7.6) at the center frequency of the specified passband to calculate l_i and d_i, which correspond to the impedance inverters. This approximate treatment of the frequency dependence of (7.1), (7.5), and (7.6) can result in a designed passband that differs considerably from that which is specified, and optimization is then required to tune the filter dimensions to satisfy the design specification.

For a given filter specification such as the two passband edge frequencies yielding λ_{gL} and λ_{gH}, passband return loss (L_R), stopband attenuation (L_I), the waveguide housing dimensions (a,b), the metal septum thickness (t), and the ridged waveguide gap (s), the modified design procedure is summarized as follows:

1. Determine the modified scaling parameter α and midband wavelength λ_{g0} from the following nonlinear equations, which can be readily solved numerically.

$$\alpha \frac{\lambda_{gL}}{\lambda_{g0}} \sin\left(\frac{\pi \lambda_{g0}}{\lambda_{gL}}\right) = 1$$

$$\alpha \frac{\lambda_{gH}}{\lambda_{g0}} \sin\left(\frac{\pi \lambda_{g0}}{\lambda_{gH}}\right) = -1 \qquad (7.7)$$

2. Determine the passband ripple level ϵ from the minimum passband return loss, which is defined as:

$$L_R = 10 \log_{10}\left(1 + \frac{1}{\epsilon^2}\right) \qquad (7.8)$$

3. Determine the number of resonators, n, from

$$L_I = 10 \log_{10}\left\{1 + \epsilon^2 \, T_n^2\left[\frac{\alpha \lambda_g}{\lambda_{g0}} \sin\left(\frac{\pi \lambda_{g0}}{\lambda_g}\right)\right]\right\} \qquad (7.9)$$

at the designated stopband frequency, f_s.

4. Calculate the impedances of the distributed element and impedance inverter values from

$$Z_r^l = \frac{2\alpha}{y} \sin\left[\frac{(2r-1)\,\pi}{2n}\right] - \frac{1}{4\alpha y}\left[\frac{-y^2 + \sin^2\left(\dfrac{r\pi}{n}\right)}{\sin\dfrac{(2r+1)\pi}{2n}}\right]$$

$$-\frac{1}{4\alpha y}\left[\frac{-y^2 + \sin^2\dfrac{2(r-1)\pi}{n}}{\sin\dfrac{(2r-3)\pi}{2n}}\right] \quad r = 1, 2, \ldots, n \quad (7.10)$$

and

$$K_{r,r+1}^l = \frac{\sqrt{y^2 + \sin^2\left(\dfrac{n\pi}{n}\right)}}{y} \quad r = 0, 1, \ldots, n \quad (7.11)$$

where

$$y = \sinh\left(\frac{1}{n}\sinh^{-1}\frac{1}{\epsilon}\right) \quad (7.12)$$

5. Recognizing that the normalized impedances of the resonators of the ridged waveguide filters (Z_r, $r = 1, \ldots, n$) are not identical to 1, we must scale the internal impedance levels of the distributed half-wave prototype filter in Figure 7.4(a) as shown in Figure 7.4(b), where

$$Z_r^* = Z_r \quad r = 0, 1, \ldots, n + 1 \quad (7.13)$$

and

$$K_{r,r+1}^* = K_{r,r+1}^l \sqrt{\frac{Z_r Z_{r+1}}{Z_r^l Z_{r+1}^l}} \quad r = 0, 1, \ldots, n \quad (7.14)$$

with

$$Z_0 = Z_{n+1} = Z_0^l = Z_{n+1}^l = 1 \quad (7.15)$$

Z_r^l and $K_{r,r+1}^l$ are the impedance values of the distributed half-wave prototype filter (see Figure 7.4) given by (7.10) and (7.11), and the

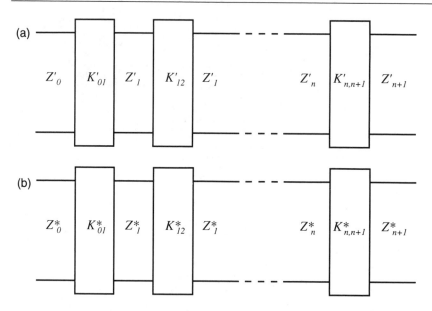

Figure 7.4 Final network (a) using impedance inverters after impedance scaling and (b) of the distributed half-wave prototype.

values of Z_r are the normalized guide impedances of the resonator sections.

6. Determine the ith septum length d_i, shown in Figure 7.5(b), by solving (7.1) so that the required impedance inverter is realized. Because those functions are not available explicitly, we must implement a root-seeking routine to find the value of width that is provided by the required normalized impedance value K for each impedance inverter.

7. Finally, the length (l_i) of the ith resonator (see Figure 7.5b) formed by the ith and $(i + 1)$th septa is given by

$$l_i = \frac{\lambda_{g0}}{2\pi}[\theta - (\phi_{2,i} + \phi_{1,i+1})] \quad i = 1, 2, \ldots, n \quad (7.16)$$

where ϕ_1 and ϕ_2 are given by (7.5) and (7.6), and the electrical distance θ corresponds to the physical distance $\lambda_{g0}/2$.

The main limitations of this approach are the frequency dependence of the guide impedances and the frequency dependence of the impedance inverters or the inaccurate approximation of the E-plane septa made in the derivation of the design procedure. Once the dimensions of the filter have been found, the frequency response of the overall filter at each frequency can be simulated by cascading the ABCD matrices of the resonators and the septa. To illustrate

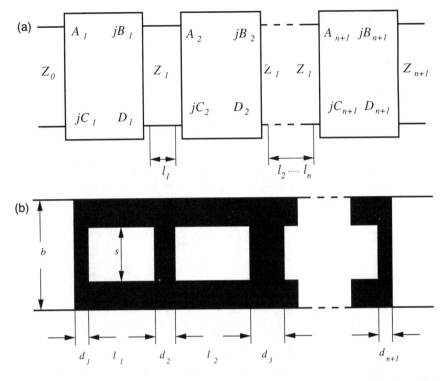

Figure 7.5 (a) Two-port representation of ridged waveguide filter using ABCD matrix for metal septum and (b) symmetrical ridged waveguide bandpass filter.

the application of this procedure to the design of E-plane bandpass filters with improved stopband performance, the design of an E-plane bandpass filter with the specifications given in Section 7.4 is considered. Figure 7.6 shows the calculated passband return loss (before optimization: the solid line) designed using this procedure. As can be seen, the design specification is still unsatisfactory, and optimization often is required in practice for the accurate design of these filters.

7.3 Numerical Implementation of Equal-Ripple Optimization

To apply the equal-ripple optimization technique described in [1] to the design of E-plane filters with improved stopband performance, it is necessary, for a given set of insert dimensions, to be able to calculate the insertion loss on a sample of frequency points within the specified passband. For a symmetrical E-plane bandpass filter with improved stopband performance, the insertion loss can be expressed in terms of an ABCD matrix. The matrix representation of the whole filter (see Figure 7.5) is

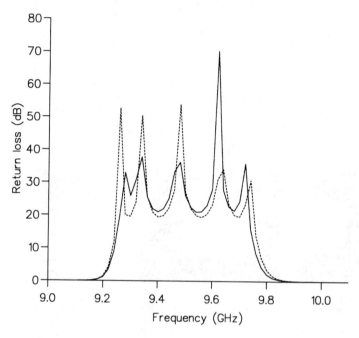

Figure 7.6 Calculated return loss before (solid line) and after (dashed line) optimization (rwg gap = 8 mm).

$$\begin{pmatrix} A & jB \\ jC & D \end{pmatrix} = \begin{pmatrix} A_1 & jB_1 \\ jC_1 & D_1 \end{pmatrix} \begin{pmatrix} \cos\theta_1 & j\dfrac{f}{\sqrt{f^2 - f_{c1}^2}}\sin\theta_1 \\ j\dfrac{\sqrt{f^2 - f_{c1}^2}}{f}\sin\theta_1 & \cos\theta_1 \end{pmatrix}$$

$$\cdot \begin{pmatrix} A_2 & jB_2 \\ jC_2 & D_2 \end{pmatrix} \begin{pmatrix} \cos\theta_2 & j\dfrac{f}{\sqrt{f^2 - f_{c2}^2}}\sin\theta_2 \\ j\dfrac{\sqrt{f^2 - f_{c2}^2}}{f}\sin\theta_2 & \cos\theta_2 \end{pmatrix}$$

$$\cdot \begin{pmatrix} \cos\theta_n & j\dfrac{f}{\sqrt{f^2 - f_{cn}^2}}\sin\theta_n \\ j\dfrac{\sqrt{f^2 - f_{c2n}^2}}{f}\sin\theta_n & \cos\theta_n \end{pmatrix} \cdot \begin{pmatrix} A_{n+1} & jB_{n+1} \\ jC_{n+1} & D_{n+1} \end{pmatrix}$$

$$(7.17)$$

with

$$\theta_i = \frac{2\pi\, l_i}{\lambda_{gi}} \tag{7.18}$$

in which l_i is the ridged waveguide resonator length, λ_{gi} is the wavelength in the ridged waveguide resonator for each frequency, and f_c is the cutoff frequency in the ridged waveguide resonator. The overall filter response (insertion loss, L_I) can be expressed in terms of elements of the total ABCD matrix of the filter at each frequency (by directly combining the ABCD matrices of the individual filter sections) as

$$L_I = 20\,\log_{10}\!\left(\frac{A + B + C + D}{2}\right) \tag{7.19}$$

The elements of the ABCD matrices of the individual filter sections are calculated using the mode-matching method [13,14]. The propagation constants of the eigenmodes in ridged waveguides are related to the cutoff frequencies, which can be calculated according to [15]. The transcendental equation of the eigenvalue of the nth mode in ridged waveguide was solved numerically. However, due to the singular behavior of the magnetic field at the edges of the septa, a large number of modes need to be included in the field expansions to ensure good convergence. That is similar to the situation for the septum in a rectangular waveguide and is due to the singular behavior of the magnetic field at the edges of the septum.

Neither accurate numerically fitted closed-form expressions nor accurate design tables for the electrical parameters of the E-plane septa in terms of septum dimensions (length and thickness) and frequency are yet available. The accurate design of ridged waveguide filters thus requires the direct calculation of the electrical parameters of E-plane septa. That highlights the need in the optimized design of those filters for optimization techniques that minimize the number of calculations of the electrical parameters of E-plane septa. A good approximate design of ridged waveguide filters can be obtained by the procedure described in Section 7.2. This procedure tries to include implicitly the actual frequency dependence of the E-plane septa and results in passbands that nearly meet design specifications. It is, therefore, adopted in this chapter as a means of generating a starting point for the optimization.

7.4 Numerical and Experimental Results

To illustrate the accuracy of the developed method for the design of symmetrical and asymmetrical ridged waveguide filters, a few ridged waveguide bandpass filters in WG16 were designed. Table 7.1 shows the specifications of symmetrical ridged waveguide bandpass filter, with ridged waveguide gap (rwg gap) = 8 mm.

Figure 7.7 shows the calculated passband return loss (solid line) of ridged waveguide filter (rwg gap = 8 mm) using the approximate method described in Section 7.2 and in [16]. Mode matching with 100 modes is used in both the design and the calculations. This approximate design was used as a starting point for equal-ripple optimization. The passband return loss calculated using the insert dimensions obtained on convergence also is shown in Figure 7.7 (dashed line). Table 7.2 lists the insert dimensions before and after optimization, which took four iterations. Mode matching with 100 modes was used throughout the optimization. Figure 7.6 showed the calculated insertion loss (solid line) of the final design of a five-resonator ridged waveguide bandpass filter (with rwg gap = 8 mm) over both the X-band (8.2 to 12.4 GHz) and the Ku-band (12.4 to 18 GHz). Also included in Figure 7.6 is a plot of the measured insertion loss (dashed line) of the fabricated design. The designed filter was fabricated using a brass waveguide housing and a copper metal insert, which was realized using spark erosion. Very good agreement between theory and experiment was observed. The measurement over both the X-band and the Ku-band was made using an HP 8510C vector network analyzer in two steps.

Step 1 was the measurement of the frequency responses of the fabricated filter between 8.2 and 12.4 GHz. A full two-port calibration with short, offset short, sliding load, and through as waveguide standards was used.

Table 7.1
X-Band Five-Resonator Ridged Waveguide Bandpass Filter Specifications

Waveguide WG16 (WR90) internal dimensions: 22.86 mm by 10.16 mm
Midband frequency: 9.50 GHz
Passband: 9.25–9.75 GHz
TE_{10} cutoff frequency in rectangular waveguide: 6.556 GHz
TE_{10} cutoff frequency in ridged waveguide: 6.450 GHz
Passband return loss: 20.00 dB
Ripple level: 0.050 dB
Number of resonators: 5
Metal insert thickness: 0.100 mm
Ridged waveguide height: 8.000 mm

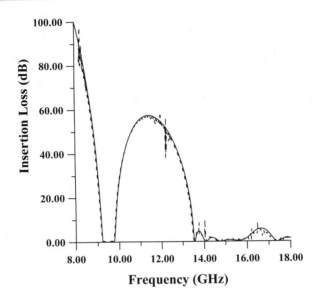

Figure 7.7 Measured (dashed line) and calculated (solid line) insertion loss of X-band five-resonator ridged waveguide bandpass filter (rwg gap = 8 mm).

Table 7.2
Symmetrical Ridged Waveguide Bandpass Filter
rwg gap = 8.00 mm
Insert thickness = 0.10 mm

| | Septum Lengths (mm) | | | Resonator Lengths (mm) | |
	Before Optimization	After Optimization		Before Optimization	After Optimization
$d_1 = d_6$:	1.3528	1.1507	$l_1 = l_5$:	15.8109	15.7789
$d_2 = d_5$:	6.4127	6.0724	$l_2 = l_4$:	16.0920	16.1070
$d_3 = d_4$:	7.7719	7.4306	l_3:	16.1032	16.1274

Step 2 was the measurement of the fabricated filter frequency response between 12 and 18 GHz. Because the cutoff frequency of the TE_{20} mode is about 13.20 GHz, that mode becomes a propagating mode for frequencies above 13.20 GHz. By using stepped WG18/WG16 transitions, the generation of the TE_{20} mode can be avoided. That is attributable to the fact that as a result of the symmetry of the transition, the TE_{20}, like all even modes, is not actually generated in the resonator waveguides. A full two-port *through-reflect-line* (TRL) calibration was used [17]. The only standards required for such calibration are a *reflect* (short), 20-mm through line, and a 25-mm *line*.

Figure 7.8 is a photograph of a five-resonator ridged waveguide bandpass filter (with rwg gap = 8 mm) together with the corresponding waveguide housing at X-band.

To demonstrate the advantages of the ridged waveguide bandpass filters over the conventional E-plane bandpass filters, a five-resonator X-band conventional E-plane bandpass filter (Table 7.3) and two ridged waveguide bandpass filters (in which the widths of the ridges are arbitrarily chosen) (Tables 7.4 and 7.5) have been designed.

Figure 7.9 shows the calculated passband return loss (solid line) of ridged waveguide filter (rwg gap = 5 mm) using the approximate method described in Section 7.2. Mode matching with 100 modes is used in both the design and the calculations. The approximate design was used as a starting point for equal-ripple optimization. The passband return loss calculated using the insert dimensions obtained on convergence (after optimization) also is shown in Figure 7.9 (dashed line). Table 7.6 lists the insert dimensions before and after optimization.

Figure 7.8 Photograph of ridged waveguide bandpass filter.

Table 7.3
Conventional E-Plane Bandpass Filter

Waveguide WG16 (WR90) internal dimension: 22.86 mm by 10.16 mm
Midband frequency: 9.5 GHz
Passband: 9.25–9.75 GHz
Number of resonators: 5
Metal insert thickness: 0.10 mm
Passband return loss: 20 dB min
Filter characteristic: Chebyshev

Table 7.4
Symmetrical Ridged Waveguide Bandpass Filter 1

Waveguide WG16 (WR90) internal dimension: 22.86 mm by 10.16 mm
Midband frequency: 9.5 GHz
Passband: 9.25–9.75 GHz
Number of resonators: 5
Metal insert thickness: 0.10 mm
Ridged waveguide gap: 8.00 mm
Passband return loss: 20 dB min
Filter characteristic: Chebyshev

Table 7.5
Symmetrical Ridged Waveguide Bandpass Filter 2

Waveguide WG16 (WR90) internal dimension: 22.86 mm by 10.16 mm
Midband frequency: 9.5 GHz
Passband: 9.25–9.75 GHz
Number of resonators: 5
Metal insert thickness: 0.10 mm
Ridged waveguide gap: 5.00 mm
Passband return loss: 20 dB min
Filter characteristic: Chebyshev

This took four iterations. Mode matching with 100 modes was used throughout the optimization. Figure 7.10 shows the comparison between calculated insertion losses of the conventional (solid line) and the ridged waveguide bandpass filter (rwg gap = 8 mm) (dashed line). Table 7.7 lists the insert dimensions for conventional E-plane and ridged waveguide bandpass filters.

Figure 7.11 shows the comparison between calculated insertion losses of the conventional (solid line) and the ridged waveguide bandpass filter

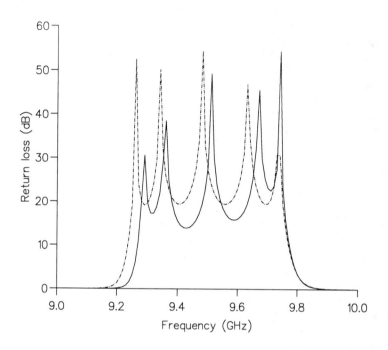

Figure 7.9 Calculated return loss before (solid line) and after (dashed line) optimization (rwg gap = 5 mm).

Table 7.6
Symmetrical Ridged Waveguide Bandpass Filter
rwg gap = 5.00 mm
Insert thickness = 0.10 mm

| | Septum Lengths (mm) | | | Resonator Lengths (mm) | |
	Before Optimization	After Optimization		Before Optimization	After Optimization
$d_1 = d_6$:	0.5615	0.4805	$l_1 = l_5$:	15.4961	15.3539
$d_2 = d_4$:	4.9200	4.6439	$l_2 = l_4$:	15.7558	15.6814
$d_3 = d_5$:	6.2612	5.9867	l_3:	15.7507	15.6829

(rwg gap = 5 mm) (dashed line). Table 7.8 lists the insert dimensions for conventional E-plane and ridged waveguide bandpass filters. As can be seen, by using ridged waveguide bandpass filters, stopband performance can be improved.

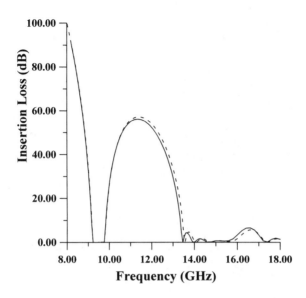

Figure 7.10 Comparison of calculated insertion loss of X-band five-resonator E-plane band-pass filters. Solid line is conventional E-plane bandpass filter; dashed line is ridged waveguide filter (rwg gap = 8 mm).

Table 7.7
Symmetrical Ridged Waveguide Bandpass Filter
rwg gap = 8.00 mm

Septum Lengths (mm)		Resonator Lengths (mm)	
Conventional Filter	**rwg Filter**	**Conventional Filter**	**rwg Filter**
$d_1 = 1.3334$	$d_1 = 1.1507$	$l_1 = 15.8258$	$l_1 = 15.7789$
$d_2 = 6.2502$	$d_2 = 6.0724$	$l_2 = 16.1814$	$l_2 = 16.1070$
$d_3 = 7.6076$	$d_3 = 7.4306$	$l_3 = 16.2054$	$l_3 = 16.1274$
$d_4 = 7.6076$	$d_4 = 7.4306$	$l_4 = 16.1814$	$l_4 = 16.1070$
$d_5 = 6.2502$	$d_5 = 6.0724$	$l_5 = 15.8258$	$l_5 = 15.7789$
$d_6 = 1.3334$	$d_6 = 1.1507$		

A two-resonator X-band asymmetrical ridged waveguide bandpass filter in WG-16 (Figure 7.12), in which the widths of the ridges are arbitrarily chosen has been designed with the specifications listed in Table 7.9.

Configuration of the asymmetrical ridged waveguide bandpass filter structure is shown in Figure 7.12. Figure 7.13 shows the calculated passband insertion loss (solid line) of the ridged waveguide filters (rwg gap 1, $s_1 = 9.00$ mm; rwg gap 2, $s_2 = 8.00$ mm) using the approximate method

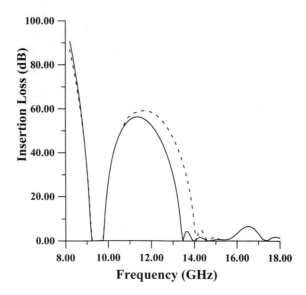

Figure 7.11 Comparison of calculated insertion loss of X-band five-resonator E-plane bandpass filters. Solid line is conventional E-plane bandpass filter; dashed line is
ridged waveguide filter (rwg gap = 5 mm).

Table 7.8
Symmetrical Ridged Waveguide Bandpass Filter
rwg gap = 5.00 mm

| Septum Lengths (mm) | | Resonator Lengths (mm) | |
Conventional Filter	rwg Filter	Conventional Filter	rwg Filter
d_1 = 1.3334	d_1 = 0.4805	l_1 = 15.8258	l_1 = 15.3539
d_2 = 6.2502	d_2 = 4.6439	l_2 = 16.1814	l_2 = 15.6814
d_3 = 7.6076	d_3 = 6.3867	l_3 = 16.2054	l_3 = 15.6829
d_4 = 7.6076	d_4 = 5.9867	l_4 = 16.1814	l_4 = 15.6814
d_5 = 6.2502	d_5 = 4.6439	l_5 = 15.8258	l_5 = 15.3539
d_6 = 1.3334	d_6 = 0.4805		

described in Section 7.2 and in [16]. The approximate design was used as a
starting point for equal-ripple optimization. Mode matching with 60 modes
was used in both the design and the calculations. Table 7.10 lists the insert
dimensions before and after optimization. The passband insertion loss calculated
using the insert dimensions obtained on convergence (after optimization) also is
shown in Figure 7.13 (dashed line). This took five Newton-Raphson iterations.
Mode matching with 60 modes was used throughout the optimization.

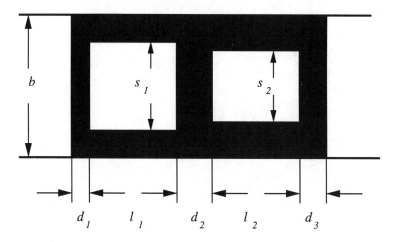

Figure 7.12 Configuration of the asymmetrical ridged waveguide bandpass filter structure.

Table 7.9
Asymmetrical Ridged Waveguide Bandpass Filter

Waveguide WG16 (WR90) internal dimension: 22.86 mm by 10.16 mm Midband frequency: 9.5 GHz Passband: 9.25–9.75 GHz Number of resonators: 2 Metal insert thickness: 0.10 mm Ridged waveguide gap 1: 9.00 mm Ridged waveguide gap 2: 8.00 mm Passband return loss: 20 dB min Filter characteristic: Chebyshev

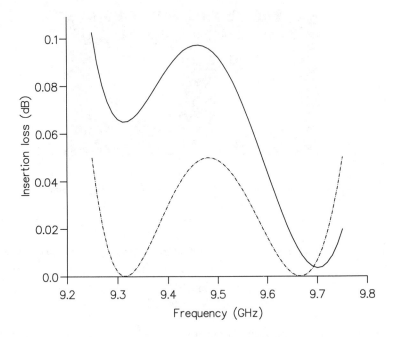

Figure 7.13 Calculated insertion loss before (solid line) and after (dashed line) optimization of the asymmetrical ridged waveguide bandpass filter.

Table 7.10
Asymmetrical Ridged Waveguide Bandpass Filter

	Septum Lengths (mm) Before Optimization	After Optimization		Resonator Lengths (mm) Before Optimization	After Optimization
d_1:	0.7458	0.8659	l_1:	15.5748	15.5169
d_2:	3.0016	3.2400	l_2:	15.3952	15.4829
d_3:	0.6003	0.6132			

References

[1] Postoyalko, V., and D. Budimir, "Design of Waveguide E-Plane Filters With All-Metal Inserts by Equal-Ripple Optimization," *IEEE Trans. Microwave Theory & Tech.*, Vol. MTT-42, February 1994, pp. 217–222.

[2] Arndt, F., "The Status of Rigorous Design of Millimeter Wave Low Insertion Loss Fin-Line and Metallic E-Plane Filters," *J. Instn. Electronics and Telecom. Engrs.*, Vol. 34, No. 2, 1988, pp. 107–119.

[3] Gololobov, V. P., and M. Yu. Omel'yanenko, "Bandpass Filters Based on Planar Metal-Dielectric Structures in the E-Plane of a Rectangular Waveguide (A Review)," *Radioelec. & Comm. Sys.*, Vol. 30, No. 1, 1987, pp. 1–15.

[4] Vahldieck, R., and W. J. R. Hoefer, "Finline and Metal Insert Filters With Improved Passband Separation and Increased Stopband Attenuation," *IEEE Trans. Microwave Theory & Tech.*, Vol. MTT-33, December 1985, pp. 1333–1339.

[5] Arndt, F., et al., "E-Plane Integrated Circuit Filters With Improved Stopband Attenuation," *IEEE Trans. Microwave Theory & Tech.*, Vol. MTT-32, October 1984, pp. 1391–1394.

[6] Gololobov, V. P., and M. Yu. Omel'yanenko, "Filters Based on Multilayered Metallic Structures in a Waveguide," *Soviet J. Commun. Technol. Electron.*, Vol. 33, Pt. 8, 1988, pp. 69–74.

[7] Arndt, F., et al., "Stopband Optimized E-Plane Filters With Multiple Metal Inserts of Variable Number per Coupling Elements," *IEE Proc.*, Vol. 133, Pt. H, June 1986, pp. 169–174.

[8] Riblet, H. J., "Waveguide Filters Having Nonidentical Sections Resonant at Same Fundamental Frequency and Different Harmonic Frequencies," U.S. Patent No. 3,153,208, 1964.

[9] Bornemann, J., and F. Arndt, "Metallic E-Plane Filter With Cavities of Different Cutoff Frequency," *IEE Electronics Letters*, Vol. 22, May 1986, pp. 524–525.

[10] Shih, Y. C., "Design of Waveguide E-Plane Filters With All Metal Inserts," *IEEE Trans. Microwave Theory & Tech.*, Vol. MTT-32, July 1984, pp. 695–704.

[11] Bui, L. Q., D. Ball, and T. Itoh, "Broad-Band Millimeter-Wave E-Plane Bandpass Filters," *IEEE Trans. Microwave Theory & Tech.*, Vol. MTT-32, December 1984, pp. 1655–1658.

[12] Lim, J. B., C. W. Lee, and T. Itoh, "An Accurate CAD Algorithm for E-Plane Type Bandpass Filters Using a New Passband Correction Method Combined With the Synthesis Procedures," *IEEE MTT-S Int. Microwave Symp. Dig.*, June 1990, pp. 1179–1182.

[13] Shih, Y. C., "The Mode-Matching Method," in *Numerical Techniques for Microwave and Millimeter-Wave Passive Structures*, T. Itoh, ed., New York: Wiley, 1989, pp. 592–621.

[14] Wexler, A., "Solution of Waveguide Discontinuities by Modal Analysis," *IEEE Trans. Microwave Theory & Tech.*, Vol. MTT-15, 1967, pp. 508–517.

[15] Montgomery, J. P., "On Complete Eigenvalue Solution of Ridged Waveguide," *IEEE Trans. Microwave Theory & Tech.*, Vol. MTT-19, June 1971, pp. 547–555.

[16] D. Budimir, "Optimized E-Plane Bandpass Filters With Improved Stop Band Performance," *IEEE Trans. Microwave Theory & Tech.*, Vol. MTT-45, February 1997, pp. 212–220.

[17] Hewlett-Packard Product Note 8510-8, "Applying the HP8510B TRL Calibration for Non-Coaxial Measurements," August 1987.

Selected Bibliography

Bornemann, J., and F. Arndt, "Waveguide E-Plane Triple-Insert Filter," *15th Eur. Microwave Conf. Dig.*, Paris, France, 1985, pp. 726–731.

DBFILTER Reference Manual, Tesla Communications Ltd., London, England.

Gupta, K. C., R. Gary, and R. Chadha, *Computer-Aided Design of Microwave Circuits*, Dedham, MA: Artech House, 1981.

Levy, R., "The Generalized Design Technique for Practical Distributed Reciprocal Ladder Networks," *IEEE Trans. Microwave Theory & Tech.*, Vol. MTT-21, August 1973, pp. 519–526.

Levy, R., "Theory of Direct Coupled Cavity Filters," *IEEE Trans. Microwave Theory & Tech.*, Vol. MTT-15, June 1967, pp. 340–348.

Matthaei, G., L. Young, and E. M. T. Jones, "*Microwave Filters, Impedance—Matching Networks and Coupling Structures*," Dedham, MA: Artech House, 1980.

Rhodes, J. D., *Theory of Electric Filters*, New York: Wiley, 1976.

Rhodes, J. D., "Microwave Circuit Realizations," in *Microwave Solid State Devices and Applications*, D. V. Morgan and M. J. Howes, eds., England: Peregrinus, 1980, pp. 49–57.

APPENDIX 7A: Derivation of the Impedance Inverter Equations

The *ABCD* matrix of the two-port network between two reference planes is

$$
\begin{pmatrix} A' & jB' \\ jC' & D' \end{pmatrix} = \begin{pmatrix} \cos\theta_1 & jZ_1\sin\theta_1 \\ \dfrac{j\sin\theta_1}{Z_1} & \cos\theta_1 \end{pmatrix}
$$

$$
\begin{pmatrix} A' & jB' \\ jC' & D' \end{pmatrix} \begin{pmatrix} \cos\theta_2 & jZ_2\sin\theta_2 \\ \dfrac{j\sin\theta_2}{Z_2} & \cos\theta_2 \end{pmatrix} \tag{7A.1}
$$

which, by definition, is an impedance inverter if $A' = D' = 0$, that is, if

$$
\left(A\cos\phi_2 - \frac{B}{Z_2}\sin\phi_2 \right)\cos\phi_1 -
$$

$$
Z_1\sin\phi_1\left(C\cos\phi_2 + \frac{D}{Z_2}\sin\phi_2 \right) = 0 \tag{7A.2}
$$

$$
(D\cos\phi_2 - CZ_2\sin\phi_2)\cos\phi_1 -
$$

$$
\frac{\sin\phi_2}{Z_1}(B\cos\phi_2 + AZ_2\sin\phi_2) = 0 \tag{7A.3}
$$

Defining

$$\tan \phi_1 = t_1 \quad \tan \phi_2 = t_2 \tag{7A.4}$$

and dividing (7A.2) and (7A.3) by $\cos f_1 \cos f_2$ gives

$$A - \frac{B}{Z_2}t_2 - C Z_1 t_1 - \frac{DZ_1}{Z_2}t_1 t_2 = 0 \tag{7A.5}$$

$$D - CZ_2 t_2 - \frac{B}{Z_1}t_1 - A\frac{Z_2}{Z_1}t_1 t_2 = 0 \tag{7A.6}$$

Solving (7A.5) for t_2 and substituting in (7A.6) gives

$$\left(D - \frac{B}{Z_1}t_1\right) - (CZ_2)\frac{(A - CZ_1 t_1)Z_2}{(B + DZ_1 t_1)} = 0 \tag{7A.7}$$

or

$$\left(\frac{BD}{Z_2} - ACZ_2\right)(1 - t_1^2) + \left(\frac{D^2 Z_1}{Z_2}\right)t_1 +$$

$$\left(-\frac{B^2}{Z_1 Z_2} + C^2 Z_1 Z_2 - \frac{A^2 Z_2}{Z_1}\right)t_1 = 0 \tag{7A.8}$$

Using the identity

$$\tan 2\phi = \frac{2 \tan \phi}{1 - \tan^2 \phi} \tag{7A.9}$$

then (7A.8) can be rearranged to give

$$\tan 2\phi_1 = \frac{2\left(\dfrac{BD}{Z_2} - ACZ_2\right)}{\dfrac{A^2 Z_2}{Z_1} + \dfrac{B^2}{Z_1 Z_2} - C^2 Z_1 Z_2 - \dfrac{D^2 Z_1}{Z_2}} \tag{7A.10}$$

Using the normalization given in (3) and (4), (7A.10) is then recognized as being equivalent to (5). Equation (6) is derived by interchanging A and D and subscripts 1 and 2 in (7A.10). Equation (1) is derived by noting that the insertion loss of the impedance inverter is given by

$$L_1 = 1 + \frac{1}{4}\left(\frac{K}{\sqrt{Z_1 Z_2}} - \frac{\sqrt{Z_1 Z_2}}{K}\right)^2 \qquad (7A.11)$$

which can be solved to give K as a function of L_I, resulting in (7.1).

8

Design of Coplanar Waveguide Filters by Optimization

In recent years, CPW has emerged as an alternative to microstrip line for the design of microwave and mm-wave integrated circuits [1]. This is due to the fact that CPWs offer several advantages over conventional microstrip lines for hybrid and monolithic MIC applications. End-coupled half-wavelength resonator filters employing capacitive coupling [2], broadside-coupled CPW filters [3], direct-coupled CPW filters employing inductive coupling [4], ribbon-of-brick-wall CPW bandpass filters [5], MMIC bandpass filters using parallel-coupled CPW lines [6], multilayer CPW filters using an overlapping of the lines [7], and edge-coupled CPW filters [8] have been reported in the literature.

General purpose optimization techniques based on least pth objective functions use general forms of error minimization algorithms [9], which simply force the filter transfer characteristic to be within specified constraints, whereas a filter must have a specified ripple characteristic, for example, the Chebyshev function. Usually the response of an optimizable filter is sampled at a number of equally spaced frequencies, and the error between that sampled response and the desired response is computed at each frequency. The optimization program, through an iterative process, reduces the error to a minimum, arriving at a final filter design in terms of the optimized filter parameters. These optimization techniques cannot be guaranteed to satisfy filter specifications and may even converge to a local minimum.

The approach presented here requires less frequency sampling than previous methods. This method optimizes the passband of a filter with respect to the Chebyshev (or minimax) criteria, which relates directly to the way filters are fabricated in practice. This vector procedure has several advantages over the general purpose optimization routines previously applied to the design of CPW filters. Design of a combined edge-end coupled bandpass filter by the use of a cascade of half-wave resonators with electromagnetic simulations driven indirectly by an equal ripple based optimizer was used as example.

Section 8.1 describes an approximate synthesis-based design procedure of CPW bandpass filters. Section 8.2 examines the numerical implementation of equal-ripple optimization, in the context of the design of CPW bandpass filters. Section 8.3 presents a design example.

8.1 An Approximate Synthesis-Based Design Procedure

8.1.1 Circuit Representation

The proposed filter structure in Figure 8.1 can be represented as shown in Figure 8.2 as a cascade of impedance inverters and resonators. The design of this type of filter usually is based on the design procedure described in Section 8.2.1, with the edge-coupled and end-coupled sections being related to impedance inverters. Figure 8.3 shows a two-port defined by its ABCD matrix. We assume that it is connected to lines of characteristic impedances Z_0 at its two ports.

The element value of the impedance inverter (see Figure 8.3) can be derived directly from the ABCD matrix of the edge- or end-coupled section. It is given by

$$K = \sqrt{L} - \sqrt{L - 1} \tag{8.1}$$

where

$$L = 1 + \frac{1}{4}[(A - D)^2 + (B - C)^2] \tag{8.2}$$

The reference plane locations are given by the equation

$$\tan(2\Phi) = \frac{2\,(AB - CD)}{(D^2 - A^2) + (B^2 - C^2)} \tag{8.3}$$

In practice, the electrical distance (Φ) can be realized as a negative value in the adjacent positive line length, which therefore becomes shortened in the final network.

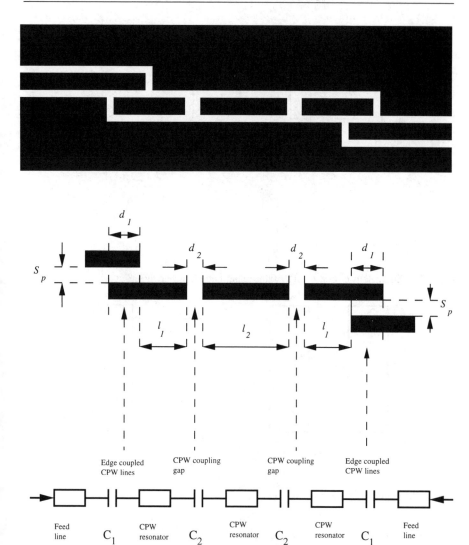

Figure 8.1 Schematic circuit diagram and layout of CPW filter.

8.1.2 Design Procedure

A common approach to the design of the conventional CPW bandpass filters [2] can be used for the filter structure described in this chapter. Here, only the most important steps in the design procedure, which include the concept of impedance inverters, are presented. The design procedure for the CPW bandpass filters is to apply (8.1) and (8.3) at the center frequency of the specified passband to calculate the θ_i and d_i, which correspond to the impedance

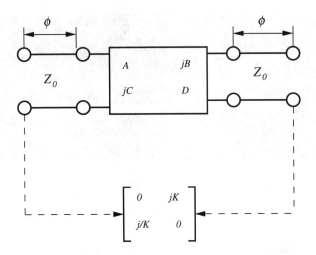

Figure 8.2 Equivalent circuit of CPW filter using impedance inverters.

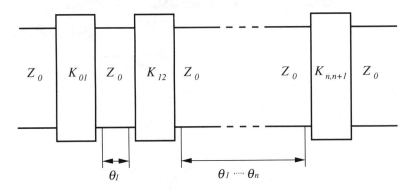

Figure 8.3 Impedance inverter.

inverters. This approximate treatment of the frequency dependence of (8.1) and (8.3) can result in a designed passband that differs considerably from that which is specified, and optimization is then required to tune the filter dimensions to satisfy the design specification.

For a given filter specification such as the two passband edge frequencies yielding f_L and f_H, passband return loss (L_R); stopband attenuation (L_I); the waveguide dimensions (Figure 8.4), such as waveguide gap (s), width (w), substrate thickness (h), the metal thickness (t); and the dielectric constant of the substrate (ϵ_r), the modified design procedure is summarized as follows:

1. Determine the passband ripple level, ϵ, from the minimum passband return loss, which is defined as

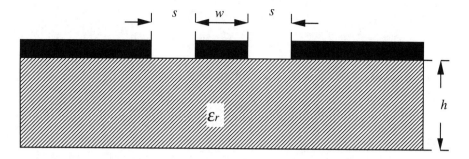

Figure 8.4 Coplanar waveguide.

$$L_R = 10 \log_{10}\left(1 + \frac{1}{\epsilon^2}\right) \tag{8.4}$$

2. Determine the number of resonators, n, from:

$$L_1 = 10 \log_{10}[1 + \epsilon^2\, T_n^2(\omega')] \tag{8.5}$$

where

$$T_n(\omega') = \cosh[n\, \text{Arcosh}(\omega')], \text{ for } |\omega'| > 1 \tag{8.6a}$$

$$T_n(\omega') = \cos[n\, \arccos(\omega')], \text{ for } 0 < \omega' \le 1 \tag{8.6b}$$

with

$$\omega' = \frac{1}{\delta}\left(\frac{f}{f_0} - \frac{f_0}{f}\right) \tag{8.7}$$

$$\delta = \frac{f_H - f_L}{f_0} \tag{8.8}$$

and

$$f_0 = \sqrt{f_H f_L} \tag{8.9}$$

at the designated stopband frequency, f_s.

3. Calculate the impedance inverter values for the first edge-coupling structure:

$$\frac{K_{0,1}}{Z_0} = \sqrt{\frac{\pi\delta}{2g_0g_1}} \tag{8.10}$$

for the intermediate end-coupling structures:

$$\frac{K_{r,r+1}}{Z_0} = \frac{\pi\delta}{2\sqrt{g_rg_{r+1}}} \quad r = 1, 2, \ldots, n-1 \tag{8.11}$$

and for the final edge-coupling structure:

$$\frac{K_{n,n+1}}{Z_0} = \sqrt{\frac{\pi\delta}{2g_ng_{n+1}}} \tag{8.12}$$

4. Determine the ith edge- or end-coupling length, d_i in Figure 8.5(a), by solving (8.1) so that the required impedance inverter corresponds to the characteristic impedance of the prototype filter. The elements of the ABCD matrix (A_i, B_i, C_i, D_i), shown in Figure 8.5(b), corresponding to the ith edge- or end-coupling section

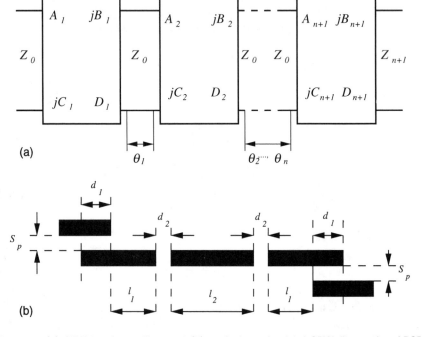

(a)

(b)

Figure 8.5 (a) CPW bandpass filter and (b) equivalent circuit of CPW filter using ABCD matrix for filter discontinuities.

(i = 1, 2, . . . , n + 1) are a function of the length of the edge-
or end-coupling section (d_i). Since those functions are not available
explicitly, we must implement a root-seeking routine to find the value
of width that is provided by the required impedance value K and the
angle Φ for each impedance inverter.

5. Finally, the length, θ_i, of the ith resonator (see Figure 8.5b) formed
   by the ith and (i + 1)th gaps is given by

$$\theta_i = \pi - (\Phi_{2,i} + \Phi_{1,i+1}) \quad i = 1, 2, \ldots, n \qquad (8.13)$$

Φ_1 and Φ_2 are given by (8.3).

Once the dimensions of the filter have been found, the frequency response
of the overall filter at each frequency can be simulated by cascading the ABCD
matrices of the resonators and the edge- and end-coupling sections. To illustrate
the application of this procedure to the design of CPW bandpass filters, the
design of an CPW bandpass filter with the specifications given in Section 8.3
(Table 8.1) is considered. Figure 8.6 shows the calculated passband return loss
(before optimization: dashed line) designed using this procedure. As can be
seen, the design specification still is unsatisfactory (e.g., passband return loss
is −4 dB at 22 GHz and −5 dB at 24 GHz; the specified value is −14 dB),
and optimization is often required in practice for the accurate design of these
filters.

8.2 Numerical Implementation of Equal-Ripple Optimization

To apply the equal-ripple optimization technique described in [10] to the design
of CPW bandpass filters, it is necessary, for a given set of filter dimensions, to
be able to calculate the insertion loss on a sample of frequency points within

Table 8.1
Coplanar Waveguide Filter Specifications

Passband: 22–24 GHz
Bandwidth: 2 GHz
Number of resonators: 3
Passband return loss: −14 dB
Connectors: CPW probes
Impedance: 50Ω

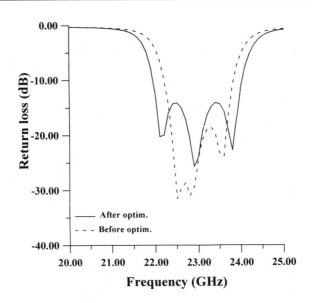

Figure 8.6 Calculated return loss before (dashed line) and after (solid line) optimization of the filter.

the specified passband. For a CPW bandpass filter, the insertion loss and the return loss can be expressed in terms of an ABCD matrix. The matrix representation of the whole filter (see Figure 8.5) is

$$\begin{pmatrix} A & jB \\ jC & D \end{pmatrix} = \begin{pmatrix} A_{dis1} & jB_{dis1} \\ jC_{dis1} & D_{dis1} \end{pmatrix} \cdot \begin{pmatrix} A_{res1} & jB_{res1} \\ jC_{res1} & D_{res1} \end{pmatrix}$$

$$\cdot \begin{pmatrix} A_{disi} & jB_{disi} \\ jC_{disi} & D_{disi} \end{pmatrix} \cdot \begin{pmatrix} A_{resi} & jB_{resi} \\ jC_{resi} & D_{resi} \end{pmatrix} \cdot \ldots \cdot \begin{pmatrix} A_{disn+1} & jB_{disn+1} \\ jC_{disn+1} & D_{disn+1} \end{pmatrix}$$

$$(8.14)$$

in which A_{dis}, B_{dis}, C_{dis} and D_{dis} are elements of the ABCD matrix of the CPW filter discontinuity such as edge- or end-coupling section and A_{res}, B_{res}, C_{res}, and D_{res} are elements of the ABCD matrix of the CPW filter resonator. The overall filter response (the insertion loss, L_I, and the return loss, L_R) can be expressed in terms of elements of the total ABCD matrix of the filter at each frequency (by directly combining the ABCD matrices of the individual filter sections) as

$$L_1 = 20 \log_{10}\left(\frac{A + B + C + D}{2}\right) \qquad (8.15)$$

$$L_R = 20 \log_{10}\left(\frac{A + B + C + D}{A + B - C - D}\right) \tag{8.16}$$

The elements of the ABCD matrices of the individual filter sections are calculated using the Em software package by Sonnet Software [11] and the Touchstone circuit simulator [12]. Neither accurate numerically fitted closed-form expressions nor accurate design tables for the electrical parameters of the edge-coupling and end-coupling sections in terms of section dimensions (length) and frequency are yet available. The accurate design of CPW filters thus requires direct calculation of the electrical parameters of those sections. That highlights the need in the optimized design of the filters for optimization techniques that minimize the number of calculations of the electrical parameters of sections. A good approximate design of a CPW filter can be obtained by that procedure, which implicitly includes the frequency dependence of the coupling gap and results in passbands that nearly meet design specifications. It is therefore adopted in this chapter as a means of generating a starting point for the optimization.

8.3 Numerical and Experimental Results

To illustrate the new approach, a three-resonator combined edge- and end-coupled CPW bandpass filter has been designed with the specifications listed in Table 8.1.

The filter can be described by four parameters: gap (d_2) and lengths (l_1, l_2, d_1), as marked in Figure 8.1. We used equal-ripple optimization with l_1, l_2, d_1, and d_2 as variables for filter, while s_p was fixed at 0.074 mm. The geometry for 50Ω CPW line on 0.635-mm substrate is width of 0.25 mm and gap of 0.127 mm. The optimization variables before and after optimization are listed in Table 8.2.

Table 8.2
Coplanar Waveguide Bandpass Filter
Thickness of the substrate = 0.635 mm
Dielectric constant of the substrate = 9.9

Parameters	Before Optimization	After Optimization
d_1 (mm):	0.195	0.642
l_1 (mm):	2.455	1.982
d_2 (mm):	0.205	0.090
l_2 (mm):	2.610	2.571

Figure 8.6 shows the calculated passband return loss of both filters (dashed line) using the approximate method. The approximate design was used as a starting point for equal-ripple optimization. The passband return loss calculated using the filter dimensions obtained on convergence are also shown in Figure 8.6 (the solid line). As can be seen, the return loss at band-edge frequencies (22 and 24 GHz) is higher than the specified value of −14 dB. Thus, optimization is required to satisfy the filter specifications.

References

[1]   Holder, P. A. R., "X-Band Microwave Integrated Circuits Using Slotlines and Coplanar Waveguide," *The Radio and Electronic Engineer*, Vol. 48, No. 1/2, January/February 1978, pp. 38–42.

[2]   Williams, D. F., and S. E. Schwarz, "Design and Performance of Coplanar Waveguide Bandpass Filters," *IEEE Trans. Microwave Theory & Tech.*, Vol. MTT-31, July 1983, pp. 558–566.

[3]   Nquyen, C., "Broadside-Coupled Coplanar Waveguides and Their End-Coupled Band-Pass Filter Applications," *IEEE Trans. Microwave Theory Tech.*, Vol. MTT-40, No. 12, December 1992, pp. 2181–2189.

[4]   Everard, J. K. A., and K. K. M. Cheng, "High Performance Direct Coupled Bandpass Filters on Coplanar Waveguide," *IEEE Trans. Microwave Theory & Tech.*, Vol. MTT-41, September 1993, pp. 1568–1573.

[5]   Lin, F-L., C-W Chiu, and R-B Wu, "Coplanar Waveguide Bandpass Filters- A Ribbon-of-Brick-Wall Design," *IEEE Trans. Microwave Theory & Tech.*, Vol. MTT-43, July 1995, pp. 1589–1596.

[6]   Mernyei, F., I. Aoki, and H. Matsuura, "MMIC Bandpass Filter Using Parallel-Coupled CPW Lines," *IEE Electronics Letters*, Vol. 30, No. 22, October 1994, pp. 1862–1863.

[7]   Menzel, W., et al., "Compact Multilayer Filter Structures for Coplanar MMIC's," *IEEE Microwave & Guide Wave Lett.*, Vol. 2, December 1992, pp. 497–498.

[8]   Karacaoglu, U., et al., "An Investigation of CPW Bandpass Filters Using End-Coupled Resonators and Square Dual-Mode Rings," *25th Eur. Microwave Conf.*, Bologna, Italy, 1995, pp. 519–523.

[9]   Bandler, J. W., and S. H. Chen, "Circuit Optimization: The State of the Art," *IEEE Trans. Microwave Theory & Tech.*, Vol. MTT-36, 1988, pp. 424–443.

[10] Postoyalko, V., and D. Budimir, "Design of Waveguide E-Plane Filters With All-Metal Inserts by Equal-Ripple Optimization," *IEEE Trans. Microwave Theory & Tech.*, Vol. MTT-42, February 1994, pp. 217–222.

[11] *Em User's Manual, Vol. 1*, Release 4.0, Sonnet Software Inc., Liverpool, NY, 1996.

[12] *Touchstone Reference Manual*, Version 3.0, EEsof Inc., Westlake Village, CA, 1991.

Selected Bibliography

Cohn, S. B., "Direct-Coupled-Resonator Filters," *Proc. IRE*, Vol. 45, February 1957, pp. 187–196.

DBFILTER Reference Manual, Tesla Communications Ltd., London, England.

Gupta, K. C., et al., *Microstrip Lines and Slotlines*, 2nd ed., Norwood, MA: Artech House, 1996.

Gupta, K. C., R. Gary, and R. Chadha, *Computer-Aided Design of Microwave Circuits*, Dedham, MA: Artech House, 1981.

Hasler, M., and J. Neiryuck, *Electrical Filters*, Dedham, MA: Artech House, 1986.

Kulke, R., and I. Wolff, "Design of Passive Coplanar Filters in V-Band," *IEEE MTT-S Int. Microwave Symp. Dig.*, 1996, pp. 1647–1650.

LINMIC+ User Manual, Version 2.1, Jansen Microwave, Germany, 1989.

M/FILTER Reference Manual, Eagleware Corp., USA, 1993.

Matthaei, G., L. Young, E. M. T. Jones, *Microwave Filters, Impedance—Matching Networks and Coupling Structures*, Dedham, MA: Artech House, 1980.

MDS Reference Manual, Release 6.0, Hewlett-Packard Co., Palo Alto, CA, 1994.

Menzel, W., W. Schwab, and G. Strauss, "Investigation of Coupling Structures for Coplanar Bandpass Filters," *IEEE MTT-S Int. Microwave Symp. Dig.*, 1995, pp. 1407–1410.

OSA90/hope Reference Manual, Version 3.5, Optimization System Associates Inc., Canada, 1995.

Rayit, A. K., and N. J. McEwan, "Coplanar Waveguide Filters," *IEEE MTT-S Int. Microwave Symp. Dig.*, 1993, pp. 1317–1320.

Schwab, W., F. Boegelsack, and W. Menzel, "Multilayer Suspended Stripline and Coplanar Line Filters," *IEEE Trans. Microwave Theory & Tech.*, Vol. MTT-42, July 1994, pp. 1403–1407.

Series IV/PC Reference Manual, Version 6.0, Hewlett-Packard Co., Palo Alto, CA, 1995.

Super-Compact User's Manual, Rev. 6.5, Compact Software Inc., Paterson, NJ, 1994.

Swanson, D. G., and R. J. Forse, "An HTS End Coupled CPW Filter at 35 GHz," *IEEE MTT-S Int. Microwave Symp. Dig.*, 1994, pp. 199–202.

9

CAD Programs

Preceding chapters have described the electromagnetic simulation, optimization, and filter design by computer optimization. All numerical results presented in this book were obtained using the DBFILTER software package, which was especially developed based on the theory outlined in this book. DBFILTER has been implemented in the FORTRAN language. The current suite (DBFILTER) includes software for synthesis and design of lumped-element lowpass filters (LCFIL), synthesis and design of lumped-element lowpass filters by computer optimization (LCFILTER), synthesis and design of CPW bandpass filters (CPWFIL), synthesis and design of CPW bandpass filters by computer optimization (CPWFILTER), electromagnetic modeling of waveguide discontinuities using mode matching (WGMODEL), synthesis of E-plane waveguide bandpass filters (EPSYNFIL), analysis and design of E-plane waveguide bandpass filters (EPFIL), design of E-plane waveguide bandpass filters by computer optimization (EPFILTER), electromagnetic modeling of ridged waveguide (RWGMODEL), synthesis of ridged waveguide bandpass filters (RWGSYNFIL), analysis and design of ridged waveguide bandpass filters (RWGFIL), and design of ridged waveguide bandpass filters by computer optimization (RWGFILTER). Computer programs are supplied for use with PC-386/486/Pentium computers and VMS and UNIX workstations. This chapter presents two computer program examples (LCFIL and EPFILTER) that show the application of the approach described in this book.

9.1 The LCFILTER Program

The LCFILTER is an optimization-oriented software package for the synthesis and design of lumped-element lowpass filters (see Figure 5.10). A flow chart for the LCFILTER program is shown in Figure 9.1. The six subroutines incorporated into the main program are as follows:

- PROTYCH calculates element values for Chebyshev lumped prototype filters. It is called by the main program.
- RESCL computes return loss and insertion loss for the whole filter.
- INDUC and CAPAC calculate the ABCD parameters for the inductors and capacitors. These subroutines are called by the subroutine RESCL.
- MULT performs ABCD matrix multiplication. It is called by the subroutine RESCL.
- EROPTIM optimizes a lumped-element lowpass filter having equal-ripple insertion loss or return loss in the passband. This subroutine is called by the main program.

The source code of the LCFIL program follows.

```
*****************************************************************
*                         LCFIL
* The main program for CAD of Lumped element lowpass filters
*
*****************************************************************
PARAMETER (M1=200,N1=200,M=5000,N=12)
INTEGER I1,II,I,I0,M0,N0,IA,IB,IJOB,IER,K0,J,J0
INTEGER NOPT,NTH,NFP,M00,N00,NM,IFIL,IRESP,ITECH,IFE
REAL*8 DET,SIGMA,F0,FS,FOP,FC,A0,B0,C0,RET,INL
REAL*8 DAVG,SUMA,DELTA,TLI,THI,Tn,INLSB,RATIO,FSTOP
REAL*8 BW,RL,EPS,RIPPLE,FH,FL,TL,TH,W,S,H,T,ER,TAND
REAL*8 LM1,LM2,LM3,LM4,LM5,LM6,LR1,LR2,LR3,LR4,LR5,LR6
REAL*8 D1,D2,D3,D4,D5,D6,D7,D8,D9,D10,D11,D12
REAL*8
AL(N+1),AE(N),AE1(N),AE2(N),AE3(N),AE4(N),AE5(N),AE6(N)
REAL*8 AJ1(N),AJ2(N),AJ3(N),AJ4(N),AJ(N,N),KINV(N+1),CAP(N+1)
REAL*8 WA(N),TC(N),AIJ(N,N),UNIT(N,N),SG(N)
REAL*8 AJT(N,N),TRM(M),REF(M),FRE(M),ROM(M),DER(N)
REAL*8 TAC(N,N),TCN(N,N),TOC(N,N),AB(N),AJ5(N),AJ6(N)
REAL*8 TRMS(M),RP(M),DIM(N+1),AERR(M),TN
REAL*8 AJ7(N),AJ8(N),AJ9(N),AJ10(N),AJ11(N),AJ12(N)
REAL*8 AE7(N),AE8(N),AE9(N),AE10(N),AE11(N),AE12(N)
REAL*8 X,Y,KK(20),LV(20),CV(20),GK(20),A0(20),BB(20),PI
REAL*8 FL,FH,Fo,Bw,Bwn,Lrip,Er,Zo,Zr,eps,ripple
REAL*8 Lr(20),Ca(20),Lbps(20),Lbpp(20),Cbps(20),Cbpp(20)
REAL*8 Lbss(20),Lbsp(20),Cbss(20),Cbsp(20)
REAL*8 Lhp(20),Chp(20),GG(20),KF(20)
CL0=299.7925000D00
      PI=4*DATAN(1.0D00)
```

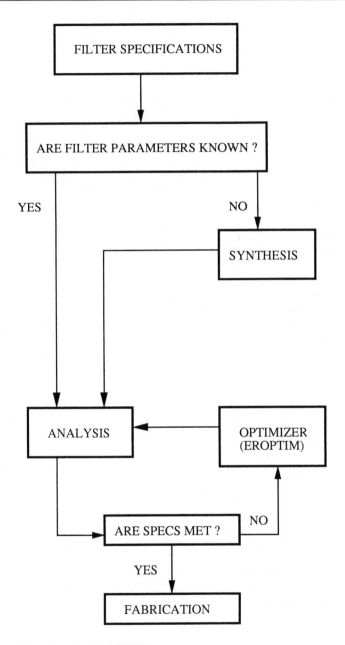

Figure 9.1 Flow chart for the LCFILTER program.

```
        ZJ=(0,1)
*       OPEN (UNIT=1,FILE='drfdim.in')
*       OPEN (UNIT=2,FILE='drf.in')
        OPEN (UNIT=6,FILE='LCFILTER.DAT')
        print *,'Number of Elements'
        read(*,*)NTH
        print *,'CUTOFF FREQUENCY in GHz '
        read(*,*)FC
        print *,' MIN. PASSBAND RETURN LOSS in dB:'
        read(*,*)RL
        print *,' MIN. STOPBAND INSERTION LOSS in dB'
        read(*,*)INLSB
        print *,' STOPBAND FREQQUENCY in GHz'
        read(*,*)Fstop
        print *,'OPERATION FREQUENCY in GHz'
        read(*,*)FOP
        print *,'FREQUENCY STEP in GHz'
        read(*,*)FSTEP
        print *,'NUMBER OF FRQUENCY POINTS'
        read(*,*)NFP
        print *,'Source Impedance'
        read(*,*)Zo
        print *,'Load Impedance'
        read(*,*)Zr
         print   *,'  Do you know filter element values YES(1)/
NO(0)'
        print   *
        print   *
        read (*,*) I
        print   *
        IF (I.EQ.1) goto 3
        goto 4
3       print *,'L1'
        read(*,*)lm1
        print *,'C1'
        read(*,*)lR1
        if(nth.eq.1) goto 50
        print *,'L2'
        read(*,*)lm2
        print *,'C2'
        read(*,*)lR2
        if(nth.eq.2) goto 50
        print *,'L3'
        read(*,*)lm3
        print *,'C3'
        read(*,*)lR3
        if(nth.eq.3) goto 50
        print *,'L4'
```

```
      read(*,*)lm4
      print *,'C4'
      read(*,*)lR4
      if(nth.eq.4) goto 50
      print *,'L5'
      read(*,*)lm5
      print *,'C5'
      read(*,*)lr5
      if(nth.eq.5) goto 50
      print *,'L6'
      read(*,*)lm6
      print *,'C6'
      read(*,*)lR6
      if(nth.eq.6) goto 50
      print *,'L7'
      read(*,*)lm7
      print *,'C7'
      read(*,*)lr7
      if(nth.eq.7) goto 50
      print *,'L8'
      read(*,*)lm8
      print *,'C8'
      read(*,*)lR8
      if(nth.eq.8) goto 50
      print *,'L9'
      read(*,*)lm9
      print *,'C9'
      read(*,*)lR9
      if(nth.eq.9) goto 50
      print *,'L10'
      read(*,*)lm10
      print *,'C10'
      read(*,*)LR10
      if(nth.eq.10) goto 50
      print *,'L11'
      read(*,*)lm11
      print *,'C11'
      read(*,*)lr11
      if(nth.eq.11) goto 50
      goto 1000
4     Fo=Fc
      BW=Fc
      BWN=BW/Fo
      EPS=DSQRT(10**(0.1*RL)-1.0)
      EPS=1/EPS
      RIPPLE=10*DLOG10(1+EPS**2)
      RATIO=FSTOP/FC
      Zn=Zr/Zo
```

```
         FS=(FL-FH)/(NFP-1)
         DO I=1,100
         EPS=(10**(0.1*RIPPLE)-1.0)
         IF (RATIO.LT.1) GOTO 6
         TN=(I*DLOG(RATIO+DSQRT(RATIO**2-1.00)))
         TN=(DEXP(TN)+DEXP(-TN))/2.00
         GOTO 7
6        TN=DCOS(I*DACOS(TN))
7        INL=10*DLOG(1+EPS**2*TN**2)
         NRES=I
         IF (NTH.GT.11) goto 1000
         IF (INL.GT.INLSB) goto 53
         ENDDO
53       EPS=10**(0.1*RIPPLE)-1.0
         RL=10*DLOG10(1.00+(1.0/EPS))
         CALL PRTYPCH (NTH,Fc,Zo,RL,LV,CV,Lr,Ca,GK,KK)
         goto 120
50       EPS=DSQRT(10**(0.1*RL)-1.0)
         EPS=1/EPS
         RIPPLE=10*DLOG10(1+EPS**2)
         LR(1)=LM1
         CA(1)=LR1
         if(nres.eq.1) goto 120
         LR(2)=LM2
         CA(2)=LR2
         if(nres.eq.2) goto 120
         LR(3)=LM3
         CA(3)=LR3
         if(nres.eq.3) goto 120
         LR(4)=LM4
         CA(4)=LR4
         if(nres.eq.4) goto 120
         LR(5)=LM5
         CA(5)=LR5
         if(nres.eq.5) goto 120
         LR(6)=LM6
         CA(6)=LR6
         if(nres.eq.6) goto 120
         CA(7)=LR7
         LR(7)=LM7
         if(nres.eq.7) goto 120
         LR(8)=LM8
         CA(8)=LR8
         if(nres.eq.8) goto 120
         CA(9)=LR9
         LR(9)=LM9
         if(nres.eq.9) goto 120
         LR(10)=LM10
```

```
      CA(10)=LR10
      if(nres.eq.10) goto 120
      CA(11)=LR11
      LR(11)=LM11
      if(nres.le.11) goto 120
      goto 1000

120   write(6,*)
      write(6,*) 'TYPICAL PERFORMANCE DATA FOR A Nth-DEGREE'
      write(6,*)
      write(6,*) '******LUMPED ELEMENT LOW PASS FILTER***'
      write(6,*)
      write(6,*)
      write(6,*)
      WRITE(6,*) 'FILTER CHARACTERISTIC:CHEBYSHEV'
      write(6,*)
      write(6,*) 'NUMBER OF ELEMENTS',NTH
      write(6,*)
      WRITE(6,*) 'CUTOFF FREQUENCY in GHz:',FC
      WRITE(6,*)
      WRITE(6,*) 'RIPPLE LEVEL IN dB',RIPPLE
      write(6,*)
      write(6,*) 'MIN. PASSBAND RETURN LOSS in dB:',RL
      WRITE(6,*)
      write(6,*) 'MIN. STOPBAND INSERTION LOSS in dB:',INLSB
      WRITE(6,*)
      write(6,*) 'STOPBAND FREQUENCY in GHz:',FSTOP
      write(6,*)
      write(6,*) 'STARTING OPERATION FREQUENCY in GHz',FOP
      write(6,*)
      write(6,*) 'NUMBER OF FREQUENCY POINTS',NFP
      write(6,*)
      write(6,*) 'FREQUENCY STEP in GHz',FS
      write(6,*)
      write(6,*)
      write(6,*) 'FILTER ELEMENT VALUES'
      write(6,*)
      write(6,*)
      write(6,*) 'LOWPASS FILTER PROTOTYPE ELEMENT VALUES'
      write(6,*)            DO I=1,NTH+2
      write(6,*) 'Gv',GK(I)
      ENDDO
      write(6,*)
      write(6,*)
      DO I=1,NTH
      write(6,*) 'Lr=Zo*gv(1/fc)',Lr(I)
      ENDDO
      write(6,*)
```

```
      DO I=1,NTH
      write(6,*) 'Ca=gv/(Zo*fc)',Ca(I)
      ENDDO
      write(6,*)

      F0=FOP
      N00=NTH+1
      M00=NFP
      print  *,'For Shunt Capacitor as 1st filter element (2)'
      print  *
      DO J=1,M00
      F0=F0+FS
      CALL RESCL(NRES,M,LR,CA,AE,F0,Zo,RET,INL)
      TRM(J)=-INL
      ROM(J)=-RET
      FRE(J)=F0
      ENDDO
      write(6,*)
       WRITE(6,*) 'FREQUENCY, RETURN and INSERTION LOSS BEFORE
OPT.'
      write(6,*)
      DO I=1,M00
      RP(I)=DELTA
      WRITE(6,*) FRE(I),ROM(I),TRM(I)
      ENDDO
1000 END
**************************************************************
SUBROUTINE PRTYPCH (NRES,Fc,Zin,RL,LV,CV,LA,CAP,GK,KK)

INTEGER Nres,Neven
REAL*8 Nodd,Ndif
REAL*8 CL0,PI,RL,RIPPLE,X0,Y0,LA(20),CAP(20),FC,Zin,GG(20)
REAL*8 X,Y,EPS,KK(20),LV(20),CV(20),GK(20),A0(20),BB(20)

CL0=299.792500D00
      PI=4*DATAN(1.0D00)

EPS=DSQRT(10**(0.1*RL)-1.0)
      EPS=1/EPS
      X0=DLOG((1/eps)+dsqrt(1+(1/eps)**2))
      X=0.5*X0
      Y0=X0/Nres
      Y=0.5*(DEXP(Y0)-DEXP(-Y0))
      DO I=1,Nres
      A0(I)=DSIN((2*I-1)*PI/(2*Nres))
      BB(I)=Y**2+(DSIN(I*PI/Nres))**2
      ENDDO
      GG(0)=1.0
```

```
          GG(1)=2.00*A0(1)/Y
          Do J=2,Nres
          GG(J)=4.00*A0(J-1)*A0(J)/(BB(J-1)*GG(J-1))
          ENDDO
          DO J=3,Nres+1
          GK(J)=GG(J-1)
          ENDDO
          GK(2)=GG(1)
          GK(1)=GG(0)
          Nodd=Nres/2.00
          Neven=Nres/2
          Ndif=Nodd-Neven
          IF(Ndif.GT.0.2) goto 1
*         Nres even
          GK(Nres+2)=((DEXP(X)+DEXP(-X))/(DEXP(X)-DEXP(-X)))**2
          goto 2
*         Nres odd 1
1         GK(Nres+2)=1.000
2         DO I=2,NRES+1
          LV(I-1)=GK(I)
          La(I-1)=LV(I-1)*Zin/(2.0*pi*Fc*1.E+9)
          ENDDO
          DO I=1,NRES
          CV(I)=GK(I+1)
          Cap(I)=CV(I)/(2.0*pi*Fc*1.E+9*Zin)
          ENDDO
          GOTO 100
3         DO I=1,NRES-1
          KK(I)=DSQRT(Y**2+(DSIN(I*PI/Nres))**2)/Y
          ENDDO
          DO I=1,NRES
          LV(I)=(2.00/Y)*DSIN((2*I-1)*PI/(2*NRES))
          La(I)=LV(I)*Zin/(2.0*pi*Fc*1.E+9)
          ENDDO
          DO I=1,NRES-1
          CV(I)=LV(I)/(KK(I))**2
          Cap(I)=CV(I)/(2.0*pi*Fc*1.E+9*Zin)
          ENDDO
1000      END
*****************************************************************
          SUBROUTINE RESCL(No,M,LR,CAP,AER,Freq,Zo,RL,IL)

          INTEGER Fpts,No,Nres,M
          COMPLEX*16 A1,B1,C1,D1,A2,B2,C2,D2,ATi,BTi,CTi,DTi
          COMPLEX*16 DET,V1,I1,Zin,ZJ,Det1,Det2,At,Bt,Ct,Dt,S11,S21
          COMPLEX*16 Att,Btt,Ctt,Dtt,At1,Bt1,Ct1,Dt1,At2,Bt2,Ct2,Dt2
          COMPLEX*16 At3,Bt3,Ct3,Dt3,At4,Bt4,Ct4,Dt4,A11,B11,C11,D11
          COMPLEX*16 A12,B12,C12,D12,A13,B13,C13,D13,A14,B14,C14,D14
```

```
COMPLEX*16 A15,B15,C15,D15,At5,Bt5,Ct5,Dt5,A3,B3,C3,D3
REAL*8 Fstart,Fstop,Er,Qu,Zo,Zn,Fo,Bw,Lrip,LP,IL,RL
REAL*8 I1M,V1M,Io,Zr,Freq,CL0,PI,S21M,S11M,S11DB,S21DB
REAL*8 LR(20),CAP(20)

*       For Nres=5 *
        REAL*8 LR(5),CAP(5),Results(3,10),G(7),K(6)
*       For Nres=6
*       REAL*8 LR(6),CAP(6),Results(3,10),G(8),K(7)
*       For Nres=7
*       REAL*8 LR(7),CAP(7),Results(3,10),G(9),K(8)
*       COMMON Er,Qu,Zo,Zn,CL0,PI,ZJ

  CL0=299.792500D00
        PI=4*DATAN(1.0D00)
        ZJ=(0,1)

*       cap(1)=2.55E-12
*       cap(3)=2.10E-12
*       cap(5)=0.49E-12
*       Lr(2)=6.25E-09
*       Lr(4)=3.66E-09

CALL CAPAC (Cap(1),Freq,Zo,A2,B2,C2,D2)
CALL INDUC (Lr(2),Freq,Zo,A1,B1,C1,D1)
CALL MULT(A2,B2,C2,D2,A1,B1,C1,D1,At,Bt,Ct,Dt)
IF(No.eq.2) GOTO 600
CALL CAPAC (Cap(3),Freq,Zo,A3,B3,C3,D3)
CALL MULT(At,Bt,Ct,Dt,A3,B3,C3,D3,At1,Bt1,Ct1,Dt1)
IF(No.eq.3) GOTO 500
CALL INDUC (Lr(4),Freq,Zo,A11,B11,C11,D11)
CALL MULT(At1,Bt1,Ct1,Dt1,A11,B11,C11,D11,At,Bt,Ct,Dt)
IF(No.eq.4) GOTO 600
CALL CAPAC (Cap(5),Freq,Zo,A2,B2,C2,D2)
CALL MULT(At,Bt,Ct,Dt,A2,B2,C2,D2,At1,Bt1,Ct1,Dt1)
IF(No.eq.5) GOTO 500
CALL INDUC (Lr(6),Freq,Zo,A11,B11,C11,D11)
CALL MULT(A11,B11,C11,D11,At1,Bt1,Ct1,Dt1,At,Bt,Ct,Dt)
IF(No.eq.6) GOTO 600
CALL CAPAC (Cap(7),Freq,Zo,A2,B2,C2,D2)
CALL MULT(A2,B2,C2,D2,At,Bt,Ct,Dt,At1,Bt1,Ct1,Dt1)
IF(No.eq.7) GOTO 500
CALL INDUC (Lr(8),Freq,Zo,A11,B11,C11,D11)
CALL MULT(A11,B11,C11,D11,At1,Bt1,Ct1,Dt1,At,Bt,Ct,Dt)
IF(No.eq.8) GOTO 600
CALL CAPAC (Cap(9),Freq,Zo,A2,B2,C2,D2)
CALL MULT(A2,B2,C2,D2,At,Bt,Ct,Dt,At1,Bt1,Ct1,Dt1)
IF(No.eq.9) GOTO 500
```

```
CALL INDUC (Lr(10),Freq,Zo,A11,B11,C11,D11)
CALL MULT(A11,B11,C11,D11,At1,Bt1,Ct1,Dt1,At,Bt,Ct,Dt)
IF(No.eq.10) GOTO 600
CALL CAPAC (Cap(11),Freq,Zo,A2,B2,C2,D2)
CALL MULT(A2,B2,C2,D2,At,Bt,Ct,Dt,At1,Bt1,Ct1,Dt1)
IF(No.eq.11) GOTO 500
CALL INDUC (Lr(12),Freq,Zo,A11,B11,C11,D11)
CALL MULT(A11,B11,C11,D11,At1,Bt1,Ct1,Dt1,At,Bt,Ct,Dt)
IF(No.eq.12) GOTO 600
CALL CAPAC (Cap(13),Freq,Zo,A2,B2,C2,D2)
CALL MULT(A2,B2,C2,D2,At,Bt,Ct,Dt,At1,Bt1,Ct1,Dt1)
IF(No.eq.13) GOTO 500
CALL INDUC (Lr(14),Freq,Zo,A11,B11,C11,D11)
CALL MULT(A11,B11,C11,D11,At1,Bt1,Ct1,Dt1,At,Bt,Ct,Dt)
IF(No.eq.14) GOTO 600
CALL CAPAC (Cap(15),Freq,Zo,A2,B2,C2,D2)
CALL MULT(A2,B2,C2,D2,At,Bt,Ct,Dt,At1,Bt1,Ct1,Dt1)
IF(No.eq.15) GOTO 500
CALL INDUC (Lr(16),Freq,Zo,A2,B2,C2,D2)
CALL MULT(A2,B2,C2,D2,At1,Bt1,Ct1,Dt1,At,Bt,Ct,Dt)
IF(No.eq.16) GOTO 600
GOTO 1000


600   S21=(2/(At+Bt+Ct+Dt))
      S21M=CDABS(S21)
      S11=((At+Bt-Ct-Dt)/(At+Bt+Ct+Dt))
      S11M=CDABS(S11)
      S21DB=20*DLOG10(S21M)
      S11DB=20*DLOG10(S11M)
      IL=20*DLOG10(1/S21M)
      RL=20*DLOG10(1/S11M)
      write(6,*)
      write(6,*) 'Frequency (GHz),Transm.,Reflec.'
      write(6,*)   Freq,S21,S11
      write(6,*) 'Frequency (GHz),Transm.(dB),Reflec.(dB)'
      write(6,*)   Freq,S21DB,S11DB
      write(6,*) 'Frequency (GHz),Ins.loss (dB),Ref.loss (dB)'
      write(6,*)   Freq,IL,RL
      write(6,*)
      goto 1000
500   S21=(2/(At1+Bt1+Ct1+Dt1))
      S21M=CDABS(S21)
      S11=((At1+Bt1-Ct1-Dt1)/(At1+Bt1+Ct1+Dt1))
      S11M=CDABS(S11)
      S21DB=20*DLOG10(S21M)
      S11DB=20*DLOG10(S11M)
      IL=20*DLOG10(1/S21M)
      RL=20*DLOG10(1/S11M)
```

```
*       write(6,*)
        write(6,*)  'Frequency (GHz),Transm.,Reflec.'
*       write(6,*)   Freq,S21,S11
        write(6,*)   Freq,S21DB,S11DB
        write(6,*)  'Frequency (GHz),Ins.loss (dB),Ref.loss (dB)'
        write(6,*)   Freq,IL,RL
        write(6,*)
1000 END
**********************************************************************
SUBROUTINE INDUC(L1,FREQ,Zo,A1,B1,C1,D1)
COMPLEX*16 GAMMA,A1,B1,C1,D1,A3,B3,C3,D3,ZJ
COMPLEX*16 GAMMA,A2,B2,C2,D2,A,B,C,D
REAL*8 Er,Qu,Zo,Zn,CL0,PI,Freq,Bw,No,Lt,L1,L2,CAP,IL
COMPLEX*16 1M,B1M,C1M,D1M,A3M,B3M,C3M,D3M,AM,DM,BM,CM
*       COMMON Er,Qu,Zo,Zn,CL0,PI,ZJ
        ZJ=(0,1)
        PI=4*DATAN(1.0D00)
        A1=(1,0)
        B1=ZJ*(2*PI*Freq*1.E+9*L1)/Zo
        C1=(0,0)
        D1=(1,0)
1000 END
**********************************************************************
SUBROUTINE CAPAC(CAP,FREQ,Zo,A2,B2,C2,D2)

COMPLEX*16 GAMMA,A1,B1,C1,D1,A3,B3,C3,D3,ZJ
COMPLEX*16 GAMMA,A2,B2,C2,D2,A,B,C,D
REAL*8 Er,Qu,Zo,Zn,CL0,PI,Freq,Bw,No,Lt,L1,L2,CAP,IL
COMPLEX*16 1M,B1M,C1M,D1M,A3M,B3M,C3M,D3M,AM,DM,BM,CM
*       COMMON Er,Qu,Zo,Zn,CL0,PI,ZJ

        ZJ=(0,1)
        PI=4*DATAN(1.0D00)

        A2=(1,0)
        B2=(0,0)
        C2=ZJ*2*PI*FREQ*1.E+9*CAP*Zo
        D2=(1,0)
1000 END
**********************************************************************
SUBROUTINE MULT(A1,B1,C1,D1,A2,B2,C2,D2,A3,B3,C3,D3)

REAL*8 IL
COMPLEX*16 A3,B3,C3,D3,A2,B2,C2,D2,A1,B1,C1,D1
        A3=A1*A2+B1*C2
        B3=A1*B2+B1*D2
        C3=C1*A2+D1*C2
        D3=C1*B2+D1*D2
1000 END
**********************************************************************
```

9.2 The EPFILTER Program

In today's increasingly competitive environment, waveguide filter designs are continuously seeking to improve product reliability and decrease performance variations without applying overly stringent manufacturing tolerances. With EPFILTER, designers can predict performance parameters quickly and easily before costly prototypes are built. EPFILTER is an optimization-oriented software package for the design of E-plane waveguide filters (see Figure 6.1) tailored to microwave and mm-wave applications. A flow chart for the EPFILTER program is shown in Figure 9.2. The software supplies a full-wave solution for filter discontinuities, using the well-known mode-matching method and equal-ripple optimization technique for filter optimization. EPFILTER software substantially reduces time to market and increases circuit performance. No other software offers such a wide range of simulation, synthesis, and optimization tools. The software is designed for ease of use.

The nine subroutines incorporated into the main program are as follows:

- PROTYCH calculates the inverters' impedance value for Chebyshev distributed prototype filters. It is called by the main program.

- STARV calculates starting filter parameters, such as lengths of the metal septa and resonators.

- EROPTIM optimizes the E-plane metal insert bandpass filter having equal-ripple insertion loss or return loss in the passband. It is called by the main program.

- MODEL computes even and odd mode impedances of the metal septum inside rectangular waveguide (waveguide bifurcation) for a given frequency and dimension. It is called by the subroutines EROPTIM and STARV.

- NVMAT performs real matrix inversion. This subroutine is called by the subroutines EROPTIM and MODEL.

- RMULL performs real matrix multiplication. This subroutine is called by the subroutine EROPTIM.

- CMULL performs multiplication of complex matrices. This subroutine is called by the subroutine MODEL.

- RETINL computes return loss and insertion loss for the whole filter. It is called by the main program, the subroutine EROPTIM, and the subroutine STARV.

- ABCD performs ABCD matrix multiplication. It is called by the subroutine RETINL.

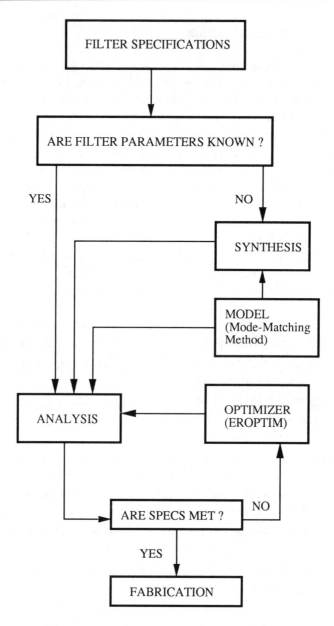

Figure 9.2 Flow chart for the EPFILTER program.

The source code of the subroutine PRTYPCH follows:

```
*****************************************************************
SUBROUTINE PRTYPCH(N,M1,N1,A0,C0,FL,FH,NFP,LI,D1S,NP,AL,
>        RL,KP,Z,ZK,Z0,ROE,ROO,TR116L,TR126L,TR216L,
>        KC1,KC,WK,INT,Ya,Yb,SUM0,SUM,SK2,Aa1,TC1,Ab1,G1,G,
>        KZ1,KZ,TC10,TCNO,TCNE,H2E,H2O,ZCO,H3NM,ZBO,ZBE,ZCE,
>        H4L,H6L,H60L,H5LM,H35LM,H30NM,H1M,UNITC,AIJC,WA,RESS)

INTEGER I,II,NFM,J,P,M1,N1,I0,IMPI,NFP,NP,NRES,N
REAL*8 D0,WA,K01,K12,K23,FC,FS,FSB,F02,TETA1,RESK01,Tn,INLSB
REAL*8 LAMDAG0,LAMDAC,LAMDA0,FI1,FI2,TETA2,FI3,FI4,RESK12
REAL*8 LAMDAL,LAMDAGL,LAMDAH,LAMDAGH,LAM1,LAM2
REAL*8 TETA3,TETA4,D5,D6,D7,D8,XS,XP,ALFA,Y,EPS,DELTA,CL0,DL0
REAL*8 Lamdas,Lamdags,FOP,FSTEP,GW0
REAL*8 Z0(N),Z(N+2),AL(N+1),KP(N+1),ZK(N+1)
REAL*8 F0,A0,B0,C0,T

        DATA  MIO,EPSO/12.566371E-10,8.85419E-15/
        DATA CL0,ETA/299.792500D00,376.70D00/
        PI=4*DATAN(1.00D00)

        NR=N
        NRES=N
        I1=M1
        II=N1
        A00=2*C0
        BW3DB=FH-FL
        KCW=PI/A0

        LAMDAC=2*PI/KCW
        LAMDAL=CL0/FL
        LAMDAH=CL0/FH
        LAMDAGL=LAMDAL/DSQRT(1-(LAMDAL/LAMDAC)**2)
        LAMDAGH=LAMDAH/DSQRT(1-(LAMDAH/LAMDAC)**2)
        LAMDAG0=(LAMDAGH+LAMDAGL)/2.0
        LAM1=LAMDAGL*DCOS(PI*LAMDAGH/(2*LAMDAGL))
        LAM2=LAMDAGH*DCOS(PI*LAMDAGL/(2*LAMDAGH))
        LAM3=DSIN(PI*LAMDAGH/(2*LAMDAGL))
        LAM4=DSIN(PI*LAMDAGL/(2*LAMDAGH))
        LAMDAG0=LAMDAG0+((LAM1+LAM2)/((LAM3+LAM4)*PI))
        ALFA=LAMDAG0/(LAMDAGL*DSIN(PI*LAMDAG0/LAMDAGL))
        LAMDA0=1+(LAMDAG0/LAMDAC)**2
        LAMDA0=DSQRT(LAMDA0)
        LAMDA0=LAMDAG0/LAMDA0
        FC=CL0/LAMDA0
        K0=2*PI/LAMDA0
        EPS=DSQRT(10**(0.1*RL)-1.0)
        EPS=1/EPS
```

```
      RIPPLE=10*DLOG10(1+EPS**2)
      EPS=DSQRT(10**(0.1*RIPPLE)-1.0)
      EPS=1/EPS
      Y=DSINH((1/NR)*DLOG(EPS+DSQRT(EPS**2+1.00)))
      EPS=1/EPS
      DO I=1,N
      Z0(I)=2*ALFA*SIN((2*I-1)*PI/(2*N))/Y-(1/(4*Y*ALFA)
>         *((Y**2+(SIN(I*PI/N))**2)/SIN((2*I+1)*PI/(2*N))
>         +(Y**2+(SIN((I-1)*PI/N))**2)/SIN((2*I-3)*PI/(2*N)))
      ENDDO
      Z(1)=1.00
      Z(N+2)=1.00
      DO J=2,N+1
      Z(J)=Z0(J-1)
      ENDDO
      ZK(1)=1.000
      DO I=2,N+1
      I0=I-1
      ZK(I)=DSQRT(1+(SIN(I0*PI/N)/Y)**2)
      ENDDO
      DO I=1,N+1
      KP(I)=ZK(I)/DSQRT(Z(I)*Z(I+1))
      ENDDO
1000  END
```

Selected Bibliography

Budimir, D., and V. Postoyalko, "EPFILTER: A CAD of Waveguide E-Plane Filters," *Microwave Journal*, August 1996.

Postoyalko, V., and D. Budimir, "Design of Waveguide E-Plane Filters With All-Metal Inserts by Equal-Ripple Optimization," *IEEE Trans. Microwave Theory & Tech.*, Vol. MTT-42, February 1994, pp. 217–222.

Budimir, D., "Design of E-Plane Filters With Improved Stopband Performance," *IEEE Trans. Microwave Theory & Tech.*, February 1997, pp. 212–220.

Appendix A Parameters

A.1 ABCD PARAMETERS

Figure A.1 shows a two-port network. ABCD parameters (V_{in}, I_{in} dependent; V_{out}, I_{out} independent) for a such network are defined as

$$\begin{bmatrix} V_{in} \\ I_{in} \end{bmatrix} = \begin{bmatrix} A & B \\ C & D \end{bmatrix} \begin{bmatrix} V_{out} \\ I_{out} \end{bmatrix} \tag{A.1}$$

For reciprocal networks:

$$AD - BC = 1 \tag{A.2}$$

For symmetrical networks:

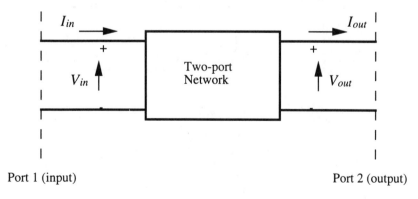

Port 1 (input) Port 2 (output)

Figure A.1 A two-port network with voltage and current defined.

$$A = D \tag{A.3}$$

For a cascade of two or more networks (Figure A.2), the overall ABCD matrix is given by

$$
\begin{bmatrix} A & B \\ C & D \end{bmatrix} = \begin{bmatrix} A_1 & B_1 \\ C_1 & D_1 \end{bmatrix} \cdot \begin{bmatrix} A_2 & B_2 \\ C_2 & D \end{bmatrix}
$$

$$
= \begin{bmatrix} A_1 A_2 + B_1 C_2 & A_1 B_2 + B_1 D_2 \\ C_1 A_2 + D_1 C_2 & C_1 B_2 + D_1 D_2 \end{bmatrix} \tag{A.4}
$$

For a parallel of two networks (Figure A.3), the overall ABCD matrix is given by

$$
\begin{bmatrix} A & B \\ C & D \end{bmatrix} = \begin{bmatrix} \dfrac{A_1 B_2 + B_1 A_2}{B_1 + B_2} & \dfrac{B_1 B_2}{B_1 + B_2} \\[4mm] \dfrac{(A_2 - A_1)(D_1 - D_2)}{B_2 + B_2} + (C_1 + C_2) & \dfrac{D_1 B_2 + B_1 D_2}{B_2 + B_2} \end{bmatrix} \tag{A.5}
$$

ABCD parameters for some of the commonly used two-port networks are shown in Figure A.4.

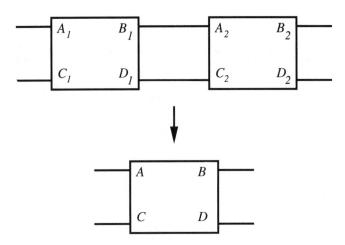

Figure A.2　A cascade connection of two-port networks.

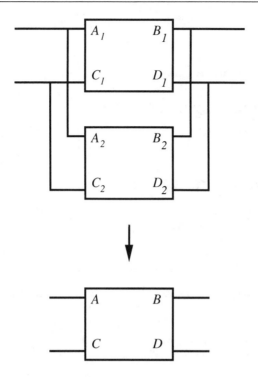

Figure A.3 A parallel connection of two-port networks.

A.2 Z Parameters

Z (impedance) parameters (V_{in}, V_{out} dependent; I_{in} I_{out} independent) for the network shown in Figure A.1 are defined as

$$\begin{bmatrix} V_{in} \\ V_{out} \end{bmatrix} = \begin{bmatrix} Z_{11} & Z_{12} \\ Z_{21} & Z_{22} \end{bmatrix} \begin{bmatrix} I_{out} \\ I_{out} \end{bmatrix} \tag{A.6}$$

A.3 Y Parameters

Y (admittance) parameters (I_{in}, I_{out} dependent; V_{out}, V_{out} independent) for the network shown in Figure A.1 are defined as

$$\begin{bmatrix} I_{in} \\ I_{out} \end{bmatrix} = \begin{bmatrix} Y_{11} & Y_{12} \\ Y_{21} & Y_{22} \end{bmatrix} \begin{bmatrix} V_{out} \\ V_{out} \end{bmatrix} \tag{A.7}$$

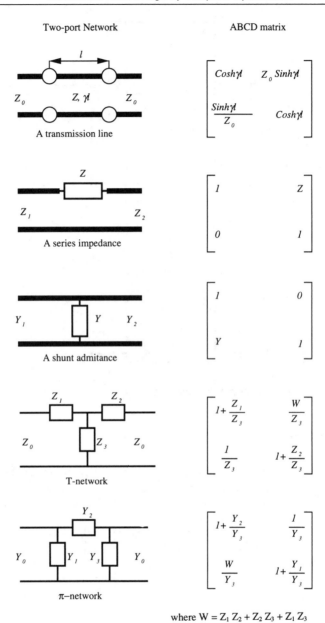

Figure A.4 A two-port network.

A.4 S Parameters

In Figure A.5, the two-port network is fed at port 1 (input) via transmission line of characteristic impedance Z_{01} and at port 2 (output) via transmission line of characteristic impedance Z_{02}. S parameters (b_{in}, b_{out} dependent; a_{in}, a_{out} independent) for such a network are defined as

$$\begin{bmatrix} b_{in} \\ b_{out} \end{bmatrix} = \begin{bmatrix} S_{11} & S_{12} \\ S_{21} & S_{22} \end{bmatrix} \begin{bmatrix} a_{in} \\ a_{out} \end{bmatrix} \tag{A.8}$$

For a reciprocal network, the S matrix is symmetrical, that is,

$$S = S^t \tag{A.9}$$

where t indicates the transpose of a matrix.

For a lossless network:

$$|S_{11}|^2 + |S_{21}|^2 = 1 \tag{A.10}$$

For a cascade of two networks with S matrices S^A and S^B (Figure A.6), the submatrices of the overall S matrix are given by

$$S_{11} = S_{11}^A + S_{12}^A S_{11}^B F S_{12}^A \tag{A.11}$$

$$S_{12} = S_{12}^A (U + S_{11}^B F S_{22}^A) S_{12}^B \tag{A.12}$$

$$S_{21} = S_{21}^B F S_{21}^A \tag{A.13}$$

$$S_{22} = S_{22}^B + S_{21}^B F S_{22}^A S_{12}^B \tag{A.14}$$

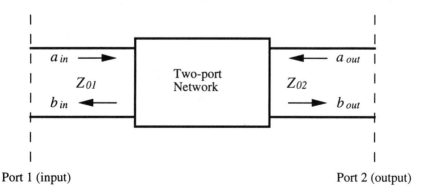

Figure A.5 A two-port network with *a* and *b* terms defined.

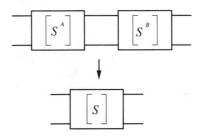

Figure A.6 A cascade connection of two S matrices.

with

$$F = (I - S_{22}^A S_{11}^B)^{-1} \tag{A.15}$$

where I is the unit matrix.

S parameters for some of the commonly used two-port networks (see Figure A.4) are as follows:

- A transmission line

$$S_{11} = \frac{(Z^2 - Z_0^2)\sinh \gamma l}{2ZZ_0 \cosh \gamma l + (Z^2 + Z_0^2)\sinh \gamma l} \tag{A.16}$$

$$S_{12} = \frac{2ZZ_0}{2ZZ_0 \cosh \gamma l + (Z^2 + Z_0^2)\sinh \gamma l} \tag{A.17}$$

$$S_{21} = \frac{2ZZ_0}{2ZZ_0 \cosh \gamma l + (Z^2 + Z_0^2)\sinh \gamma l} \tag{A.18}$$

$$S_{22} = \frac{(Z^2 - Z_0^2)\sin h\gamma}{2ZZ_0 \cosh \gamma l + (Z^2 + Z_0^2)\sinh \gamma l} \tag{A.19}$$

- A series impedance

$$S_{11} = \frac{Z - Z_1 + Z_2}{Z + Z_1 + Z_2} \tag{A.20}$$

$$S_{12} = \frac{2\sqrt{Z_1 Z_2}}{Z + Z_1 + Z_2} \tag{A.21}$$

$$S_{21} = \frac{2\sqrt{Z_1 Z_2}}{Z + Z_1 + Z_2} \tag{A.22}$$

$$S_{22} = \frac{Z + Z_1 - Z_2}{Z + Z_1 + Z_2} \tag{A.23}$$

- A shunt admittance

$$S_{11} = \frac{Y_1 - Y_2 - Y}{Y + Y_1 + Y_2} \tag{A.24}$$

$$S_{12} = \frac{\sqrt{2Y_1Y_2}}{Y + Y_1 + Y_2} \tag{A.25}$$

$$S_{21} = \frac{2\sqrt{Y_1Y_2}}{Y + Y_1 + Y_2} \tag{A.26}$$

$$S_{22} = \frac{Y_2 - Y_1 - Y}{Y + Y_1 + Y_2} \tag{A.27}$$

- A *T*-network

$$S_{11} = \frac{Z_1Z_2 + Z_2Z_3 + Z_3Z_1 + Z_0(Z_1 - Z_2) - Z_0^2}{Z_0^2 + Z_1Z_2 + Z_2Z_3 + Z_3Z_1 + Z_0(Z_1 + Z_2 + 2Z_3)} \tag{A.28}$$

$$S_{12} = \frac{2Z_0Z_2}{Z_0^2 + Z_1Z_2 + Z_2Z_3 + Z_3Z_1 + Z_0(Z_1 + Z_2 + 2Z_3)} \tag{A.29}$$

$$S_{21} = \frac{2Z_0Z_2}{Z_0^2 + Z_1Z_2 + Z_2Z_3 + Z_3Z_1 + Z_0(Z_1 + Z_2 + 2Z_3)} \tag{A.30}$$

$$S_{22} = \frac{Z_1Z_2 + Z_2Z_3 + Z_3Z_1 - Z_0(Z_1 - Z_2) - Z_0^2}{Z_0^2 + Z_1Z_2 + Z_2Z_3 + Z_3Z_1 + Z_0(Z_1 + Z_2 + 2Z_3)} \tag{A.31}$$

- A π-network

$$S_{11} = \frac{Y_0^2 - (Y_1Y_2 + Y_2Y_3 + Y_3Y) - Y_0(Y_1 - Y_2)}{Y_0^2 + (Y_1Y_2 + Y_2Y_3 + Y_3Y) + Y_0(Y_1 + Y_2 + 2Y_3)} \tag{A.32}$$

$$S_{12} = \frac{2Y_0Y_3}{Y_0^2 + (Y_1Y_2 + Y_2Y_3 + Y_3Y) + Y_0(Y_1 + Y_2 + 2Y_3)} \tag{A.33}$$

$$S_{21} = \frac{2Y_0Y_3}{Y_0^2 + (Y_1Y_2 + Y_2Y_3 + Y_3Y) + Y_0(Y_1 + Y_2 + 2Y_3)} \tag{A.34}$$

$$S_{22} = \frac{Y_0^2 - (Y_1Y_2 + Y_2Y_3 + Y_3Y) + Y_0(Y_1 - Y_2)}{Y_0^2 + (Y_1Y_2 + Y_2Y_3 + Y_3Y) + Y_0(Y_1 + Y_2 + 2Y_3)} \tag{A.35}$$

A.5 T Parameters

T parameters (a_{in}, b_{in} dependent; b_{out}, a_{out} independent) for the network shown in Figure A.5 are defined as

$$\begin{bmatrix} a_{in} \\ b_{in} \end{bmatrix} = \begin{bmatrix} T_{11} & T_{12} \\ T_{21} & T_{22} \end{bmatrix} \begin{bmatrix} b_{out} \\ a_{out} \end{bmatrix} \tag{A.36}$$

For reciprocal networks:

$$T_{11}T_{22} - T_{12}T_{21} = 1 \tag{A.37}$$

For symmetrical networks:

$$T_{21} = -T_{12} \tag{A.38}$$

For a cascade of two networks with T matrices, T^A and T^B (Figure A.7), the overall T matrix is

$$\begin{bmatrix} T_{11} & T_{12} \\ T_{21} & T_{22} \end{bmatrix} = \begin{bmatrix} T_{11}^A & T_{12}^A \\ T_{21}^A & T_{22}^A \end{bmatrix} \cdot \begin{bmatrix} T_{11}^B & T_{12}^B \\ T_{21}^B & T_{22}^B \end{bmatrix} \tag{A.39}$$

A.6 Relationships Between Z and Y Parameters

A.6.1 From Z to Y Parameters

$$Y_{11} = \frac{Z_{22}}{Z_{11}Z_{22} - Z_{12}Z_{21}} \tag{A.40}$$

$$Y_{12} = \frac{-Z_{12}}{Z_{11}Z_{22} - Z_{12}Z_{21}} \tag{A.41}$$

$$Y_{21} = \frac{-Z_{21}}{Z_{11}Z_{22} - Z_{12}Z_{21}} \tag{A.42}$$

$$Y_{22} = \frac{Z_{11}}{Z_{11}Z_{22} - Z_{12}Z_{21}} \tag{A.43}$$

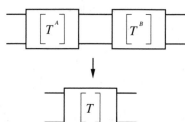

Figure A.7 A cascade connection of two T matrices.

A.6.2 From Y to Z Parameters

$$Z_{11} = \frac{Y_{22}}{Y_{11}Y_{22} - Y_{12}Y_{21}} \tag{A.44}$$

$$Z_{12} = \frac{-Y_{12}}{Y_{11}Y_{22} - Y_{12}Y_{21}} \tag{A.45}$$

$$Z_{21} = \frac{-Y_{21}}{Y_{11}Y_{22} - Y_{12}Y_{21}} \tag{A.46}$$

$$Z_{22} = \frac{Y_{11}}{Y_{11}Y_{22} - Y_{12}Y_{21}} \tag{A.47}$$

A.7 Relationships Between Z and ABCD Parameters

A.7.1 From Z to ABCD Parameters

$$A = \frac{Z_{11}}{Z_{21}} \tag{A.48}$$

$$B = \frac{Z_{11}Z_{22} - Z_{12}Z_{21}}{Z_{21}} \tag{A.49}$$

$$C = \frac{1}{Z_{21}} \tag{A.50}$$

$$D = \frac{Z_{22}}{Z_{21}} \tag{A.51}$$

A.7.2 From ABCD to Z Parameters

$$Z_{11} = \frac{A}{C} \tag{A.52}$$

$$Z_{12} = \frac{AD - BC}{C} \tag{A.53}$$

$$Z_{21} = \frac{1}{C} \tag{A.54}$$

$$Z_{22} = \frac{D}{C} \tag{A.55}$$

A.8   Relationships Between Y and ABCD Parameters

A.8.1   From Y to ABCD Parameters

$$A = \frac{-Y_{22}}{Y_{21}} \tag{A.56}$$

$$B = \frac{-1}{Y_{21}} \tag{A.57}$$

$$C = \frac{Y_{12}Y_{21} - Y_{11}Y_{22}}{Y_{21}} \tag{A.58}$$

$$D = \frac{-Y_{11}}{Y_{21}} \tag{A.59}$$

A.8.2   From ABCD to Y Parameters

$$Y_{11} = \frac{D}{B} \tag{A.60}$$

$$Y_{12} = \frac{BC - AD}{B} \tag{A.61}$$

$$Y_{21} = \frac{-1}{B} \tag{A.62}$$

$$Y_{22} = \frac{A}{B} \tag{A.63}$$

A.9   Relationships Between S and ABCD Parameters

A.9.1   From S to ABCD Parameters

$$A = \left(\frac{1 + S_{11} - S_{22} - \Delta S}{2S_{21}} \right) \sqrt{\frac{Z_{01}}{Z_{02}}} \tag{A.64}$$

$$B = \left(\frac{1 + S_{11} + S_{22} + \Delta S}{2S_{21}} \right) \sqrt{\frac{Z_{01}}{Z_{02}}} \tag{A.65}$$

$$C = \left(\frac{1 - S_{11} - S_{22} + \Delta S}{2S_{21}} \right) \sqrt{\frac{Z_{01}}{Z_{02}}} \tag{A.66}$$

$$D = \left(\frac{1 - S_{11} + S_{22} - \Delta S}{2S_{21}} \right) \sqrt{\frac{Z_{02}}{Z_{01}}} \tag{A.67}$$

where Z_{01} and Z_{02} are the normalized impedances at ports 1 (input) and 2 (output), respectively, and

$$\Delta S = S_{11}S_{22} - S_{21}S_{12} \qquad (A.68)$$

A.9.2 From ABCD to S Parameters

$$S_{11} = \left(\frac{AZ_{02} + B - CZ_{01}Z_{02} - DZ_{01}}{AZ_{02} + B + CZ_{01}Z_{02} + DZ_{01}} \right) \qquad (A.69)$$

$$S_{12} = \left(\frac{2(AD - BC)\sqrt{Z_{01}Z_{02}}}{AZ_{02} + B + CZ_{01}Z_{02} + DZ_{01}} \right) \qquad (A.70)$$

$$S_{21} = \left(\frac{2\sqrt{Z_{01}Z_{02}}}{AZ_{02} + B + CZ_{01}Z_{02} + DZ_{01}} \right) \qquad (A.71)$$

$$S_{22} = \left(\frac{-AZ_{02} + B - CZ_{01}Z_{02} + DZ_{01}}{AZ_{02} + B + CZ_{01}Z_{02} + DZ_{01}} \right) \qquad (A.72)$$

A.10   Relationships Between S and Z Parameters

A.10.1   From S to Z Parameters

$$Z_{11} = \frac{Z_{01}[(1 + S_{11})(1 - S_{22}) + S_{12}S_{21}]}{(1 - S_{11})(1 - S_{22}) - S_{12}S_{21}} \qquad (A.73)$$

$$Z_{12} = \frac{2Z_{01}S_{12}}{(1 - S_{11})(1 - S_{22}) - S_{12}S_{21}} \qquad (A.74)$$

$$Z_{21} = \frac{2Z_{02}S_{21}}{(1 - S_{11})(1 - S_{22}) - S_{12}S_{21}} \qquad (A.75)$$

$$Z_{22} = \frac{Z_{02}[(1 - S_{11})(1 + S_{22}) + S_{12}S_{21}]}{(1 - S_{11})(1 - S_{22}) - S_{12}S_{21}} \qquad (A.76)$$

A.10.2 From Z to S Parameters

$$S_{11} = \frac{\left(\dfrac{Z_{11}}{Z_{01}} - 1\right)\left(\dfrac{Z_{22}}{Z_{02}} + 1\right) - \dfrac{Z_{12}Z_{21}}{Z_{01}Z_{02}}}{\left(\dfrac{Z_{11}}{Z_{01}} + 1\right)\left(\dfrac{Z_{22}}{Z_{02}} + 1\right) - \dfrac{Z_{12}Z_{21}}{Z_{01}Z_{02}}} \tag{A.77}$$

$$S_{12} = \frac{2\dfrac{Z_{12}}{Z_{02}}}{\left(\dfrac{Z_{11}}{Z_{01}} + 1\right)\left(\dfrac{Z_{22}}{Z_{02}} + 1\right) - \dfrac{Z_{12}Z_{21}}{Z_{01}Z_{02}}} \tag{A.78}$$

$$S_{21} = \frac{2\dfrac{Z_{21}}{Z_{01}}}{\left(\dfrac{Z_{11}}{Z_{01}} + 1\right)\left(\dfrac{Z_{22}}{Z_{02}} + 1\right) - \dfrac{Z_{12}Z_{21}}{Z_{01}Z_{02}}} \tag{A.79}$$

$$S_{22} = \frac{\left(\dfrac{Z_{11}}{Z_{01}} + 1\right)\left(\dfrac{Z_{22}}{Z_{02}} - 1\right) - \dfrac{Z_{12}Z_{21}}{Z_{01}Z_{02}}}{\left(\dfrac{Z_{11}}{Z_{01}} + 1\right)\left(\dfrac{Z_{22}}{Z_{02}} + 1\right) - \dfrac{Z_{12}Z_{21}}{Z_{01}Z_{02}}} \tag{A.80}$$

A.11 Relationships Between S and Y Parameters

A.11.1 From S to Y Parameters

$$Y_{11} = \frac{Y_{01}[(1 - S_{11})(1 + S_{22}) + S_{12}S_{21}]}{(1 + S_{11})(1 + S_{22}) - S_{12}S_{21}} \tag{A.81}$$

$$Y_{12} = \frac{-2Y_{01}S_{12}}{(1 + S_{11})(1 + S_{22}) - S_{12}S_{21}} \tag{A.82}$$

$$Y_{21} = \frac{-2Y_{02}S_{21}}{(1 + S_{11})(1 + S_{22}) - S_{12}S_{21}} \tag{A.83}$$

$$Y_{22} = \frac{Y_{02}[(1 + S_{11})(1 - S_{22}) + S_{12}S_{21}]}{(1 + S_{11})(1 + S_{22}) - S_{12}S_{21}} \tag{A.84}$$

A.11.2 From Y to S Parameters

$$S_{11} = \frac{\left(1 - \dfrac{Y_{11}}{Y_{01}}\right)\left(1 + \dfrac{Y_{22}}{Y_{02}}\right) + \dfrac{Y_{12}Y_{21}}{Y_{01}Y_{02}}}{\left(1 + \dfrac{Y_{11}}{Y_{01}}\right)\left(1 + \dfrac{Y_{22}}{Y_{02}}\right) - \dfrac{Y_{12}Y_{21}}{Y_{01}Y_{02}}} \qquad (A.85)$$

$$S_{12} = \frac{-2\dfrac{Y_{12}}{Y_{01}}}{\left(1 + \dfrac{Y_{11}}{Y_{01}}\right)\left(1 + \dfrac{Y_{22}}{Y_{02}}\right) - \dfrac{Y_{12}Y_{21}}{Y_{01}Y_{02}}} \qquad (A.86)$$

$$S_{21} = \frac{-2\dfrac{Y_{21}}{Y_{02}}}{\left(1 + \dfrac{Y_{11}}{Y_{01}}\right)\left(1 + \dfrac{Y_{22}}{Y_{02}}\right) - \dfrac{Y_{12}Y_{21}}{Y_{01}Y_{02}}} \qquad (A.87)$$

$$S_{22} = \frac{\left(1 + \dfrac{Y_{11}}{Y_{01}}\right)\left(1 - \dfrac{Y_{22}}{Y_{02}}\right) + \dfrac{Y_{12}Y_{21}}{Y_{01}Y_{02}}}{\left(1 + \dfrac{Y_{11}}{Y_{01}}\right)\left(1 + \dfrac{Y_{22}}{Y_{02}}\right) - \dfrac{Y_{12}Y_{21}}{Y_{01}Y_{02}}} \qquad (A.88)$$

A.12 Relationships Between S and T Parameters

A.12.1 From T to S Parameters

$$S_{11} = \left(\frac{T_{12}}{T_{11}}\right) \qquad (A.89)$$

$$S_{12} = \left(\frac{T_{11}T_{22} - T_{12}T_{21}}{T_{22}}\right) \qquad (A.90)$$

$$S_{21} = \left(\frac{1}{T_{22}}\right) \qquad (A.91)$$

$$S_{22} = \left(\frac{-T_{21}}{T_{22}}\right) \qquad (A.92)$$

A.12.2 From S to T Parameters

$$T_{11} = \left(\frac{S_{12}S_{21} - S_{11}S_{22}}{S_{21}}\right) \tag{A.93}$$

$$T_{12} = \left(\frac{S_{11}}{S_{21}}\right) \tag{A.94}$$

$$T_{21} = \left(\frac{-S_{22}}{S_{21}}\right) \tag{A.95}$$

$$T_{22} = \left(\frac{1}{S_{21}}\right) \tag{A.96}$$

A.13 Relationships Between ABCD and T Parameters

A.13.1 From ABCD to T Parameters

$$T_{11} = \left(\frac{AZ_{02} + B + CZ_{01}Z_{02} + DZ_{01}}{2\sqrt{Z_{01}Z_{02}}}\right) \tag{A.97}$$

$$T_{12} = \left(\frac{AZ_{02} - B + CZ_{01}Z_{02} - DZ_{01}}{2\sqrt{Z_{01}Z_{02}}}\right) \tag{A.98}$$

$$T_{21} = \left(\frac{AZ_{02} + B - CZ_{01}Z_{02} - DZ_{01}}{2\sqrt{Z_{01}Z_{02}}}\right) \tag{A.99}$$

$$T_{22} = \left(\frac{AZ_{02} - B - CZ_{01}Z_{02} + DZ_{01}}{2\sqrt{Z_{01}Z_{02}}}\right) \tag{A.100}$$

A.13.2 From T to ABCD Parameters

$$A = \left(\frac{T_{11} + T_{12} + T_{21} + T_{22}}{2}\right) \sqrt{\frac{Z_{01}}{Z_{02}}} \tag{A.101}$$

$$B = \left(\frac{T_{11} - T_{12} + T_{21} - T_{22}}{2}\right) \sqrt{Z_{01}Z_{02}} \tag{A.102}$$

$$C = \frac{T_{11} + T_{12} - T_{21} - T_{22}}{2\sqrt{Z_{01}Z_{02}}} \tag{A.103}$$

$$D = \left(\frac{T_{11} - T_{21} - T_{12} + T_{22}}{2}\right) \sqrt{\frac{Z_{02}}{Z_{01}}} \tag{A.104}$$

A.14 Relationships Between S Parameters and Z_{oe} (Z_{oo}) Input Impedance

Figure A.8(a) shows a two-port passive lossless linear network. By using Bartlett's bisection theorem, the S parameters for such network can be written as

$$S_{11} = S_{22} = \left[\frac{1 - Z_{oe}Z_{oo}}{(1 + Z_{oe})(1 + Z_{oo})} \right] \qquad (A.105)$$

$$S_{21} = S_{12} = \left[\frac{Z_{oo} - Z_{oe}}{(1 + Z_{oe})(1 + Z_{oo})} \right] \qquad (A.106)$$

where Z_{oe} and Z_{oo} are the even- and odd-mode impedances of a symmetrical network. Z_{oe} (Z_{oo}) is the input impedance of the two identical one-ports formed by placing a magnetic (electric) wall at the plane of symmetry, as shown in Figure A.8(b) and Figure A.8(c), respectively, and is a reactance function.

Selected Bibliography

Bahl, I. J., and P. Bhartia, *Microwave Solid State Circuit Design*, New York: Wiley, 1988.

Gupta, K. C., R. Garg, and R. Chadha, *Computer-Aided Design of Microwave Circuits*, Norwood, MA: Artech House, 1981.

Rizzi, P. A., *Microwave Engineering Passive Circuits*, Englewood Cliffs, NJ: Prentice Hall, 1988.

Saad, T. S., *Microwave Engineers Handbook, Vol. 1*, Dedham, MA: Artech House, 1971.

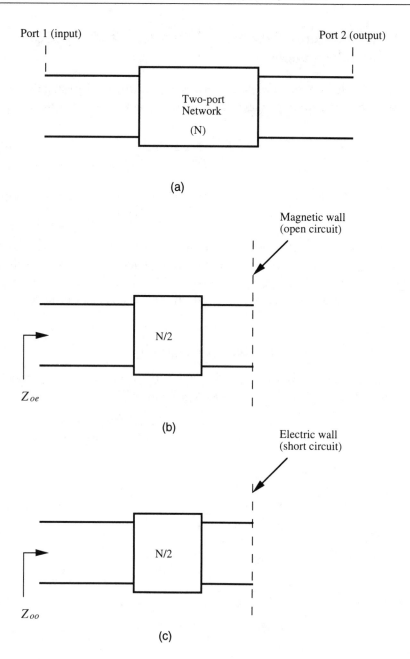

Figure A.8 (a) Two-port symmetrical network; (b) even-mode impedance; (c) odd-mode impedance.

Appendix B

B.1 Physical Constants

Velocity of light in free space, c: 2.997925×10^8 m/s
Permittivity of free space, ϵ_o: $8.854 \times 10^{-12} \approx (1/36\pi) \times 10^{-9}$ F/m
Permeability of free space, μ_o: $4\pi \times 10^{-7}$ H/m
Free-space wave impedance, η_o: $376.7 \approx 120\pi$ Ω
Boltzmann's constant, k: 1.380×10^{-23} J/K
Electron rest mass, m_0: 9.1095×10^{-31} kg
Proton rest mass: 1.67252×10^{-27} kg
Electron charge magnitude, e: 1.6022×10^{-19} C

B.2 Resistivity and Conductivity of Some Common Metals

Material	Resistivity $\rho(\mu\Omega\text{-cm})$	Conductivity $\sigma(\text{S/m})$
Silver	1.6	6.3×10^7
Copper	1.7	5.9×10^7
Gold	2.4	4.2×10^7
Aluminum	2.9	3.4×10^7
Brass	6.7	1.5×10^7
Platinum	12	8.3×10^6
Chromium	13	7.7×10^6
Tantalum	14	7.1×10^6

B.3 Properties of Common Substrate Materials at Microwave Frequencies

Material	Relative Permittivity, ϵ_r	Loss Tangent, tan δ
Silicon	11.70–12.90	0.001–0.003
Alumina	9.60–10.10	0.0005–0.002
Germanium	16.00	
Gallium arsenide	12.90	0.0005–0.001
Sapphire	9.40	0.0002
Beryllium oxide	6.70	0.001–0.002
PTFE/woven glass	2.84	0.001–0.002
PTFE/microfiberglass	2.26	0.0005–0.001
RT/Duroid 5880	2.26	0.0010
RT/Duroid 6006	6.36	
Teflon	2.10	
Epsilam 10	13.00	

List of Principal Symbols and Abbreviations

α	Scaling parameter
β	Phase constant
$\Delta(f)$	Passband correction factor
δ	A step in the calculation of the Jacobian matrix
$\tan \delta$	Loss tangent
ϵ	Passband ripple level
ϵ_0	Permittivity of free space
ϵ_r	Relative permittivity
η_0	Free-space wave impedance
γ	Propagation constant
λ_c	Cutoff wavelength
λ_g	Guide wavelength
λ_{gl}	Guide wavelength at the lower bandedge frequency
λ_{go}	Midband wavelength
λ_{gu}	Guide wavelength at the upper bandedge frequency
λ_0	Free space wavelength
μ_0	Permeability of free space
μ_r	Relative permeability
ρ	Reflection coefficient
r	Resistivity
s	Conductivity
f	Electrical length of the impedance inverter

w	Angular frequency
A, B, C, D	Elements of the ABCD matrix
a, b	Waveguide housing dimensions
ABCD	ABCD matrix
c	Velocity of light in free space
C_i	Capacitance values of the lowpass prototype filter
d_i	Length of the ith septum
$E(x_i)$	Error vector
E_t	Tangential component of electric field
E_x	x component of electric field
E_y	y component of electric field
E_z	z component of electric field
f	Operation frequency
f_0	Center frequency
f_c	Cutoff frequency
f_L	Lower bandedge frequency
f_s	Stopband frequency
f_H	Upper bandedge frequency
g_i	Normalized element values of the lowpass prototype filter
h	Thickness of the substrate
H_t	Tangential component of magnetic field
H_x	x component of magnetic field
H_y	y component of magnetic field
H_z	z component of magnetic field
I	Unity matrix
J	Jacobian matrix
k_c	Cutoff wave number
k_o	Free-space wave number
K_r	Characteristic impedance of rth impedance inverter
L_I	Insertion loss
L_i	Inductance values of the lowpass prototype filter
l_i	Length of the ith resonator
L_R	Return loss
Q_h	Magnetic hertzian potential
Q_u	Unloaded quality factor
S	Scattering matrix

$S_{11}, S_{12}, S_{21}, S_{22}$	Elements of the scattering matrix
s_i	Gap of the ith ridged waveguide
T	Transmission matrix
t	Thickness of the metal insert and the ridged waveguide
$T_{11}, T_{12}, T_{21}, T_{22}$	Elements of the T matrix
T_f	Temperature coefficient
T_n	nth degree Chebyshev polynomial of the first kind
x_i	Vector components
x_{si}, x_{pi}	Normalized reactances
Y	Admittance matrix
$Y_{11}, Y_{12}, Y_{21}, Y_{22}$	Elements of the admittance matrix
Y_i	Guide admittance of the ith mode
Z	Impedance matrix
$Z_{11}, Z_{12}, Z_{21}, Z_{22}$	Elements of the impedance matrix
z_e	Normalized even-mode impedance
Z_i	Guide impedance of the ith mode
z_o	Normalized odd-mode impedance
Z_r	Characteristic impedance of rth unite element
BPF	Bandpass filter
BSF	Bandstop filter
CAD	Computer-aided design
CAFD	Computer-aided filter design
CPU	Central processor unit
CPW	Coplanar waveguide
CPWFIL	Software for synthesis and design of coplanar waveguide filters
CPWFILTER	Software for synthesis and design of coplanar waveguide filters by computer optimization
dB	Decibel
det	Determinant
DR	Dielectric resonator
EM	Electromagnetic modeling
EPFIL	Software for analysis and design of E-plane waveguide filters

EPFILTER WGMODEL	Software for electromagnetic modelling of waveguide discontinuities
EPSYNFIL	Software for synthesis of E-plane waveguide filters
EW	Electronic warfare
GHz	Gigahertz
GPS	Global positioning system
GSM	Global system for mobile communications
HFSS	High-frequency structure simulator
HPF	Highpass filter
KCC	Kimberley Communications Consultants
LCFIL	Software for synthesis and design of lumped-element filters
LCFILTER	Software for synthesis and design of lumped-element filters by computer optimization
log	Common logarithm
LPF	Lowpass filter
MDS	Microwave design suite
MMIC	Monolithic Microwave Integrated Circuit
PCN	Personal communication network
PCS	Personal communication system
Quasi-TEM	Quasi-transverse electromagnetic
RAM	Random access memory
RF	Radio frequency
RISC	Reduced instruction set computer processor
rwg	Ridged waveguide
RWGFIL	Software for analysis and design of ridged waveguide filters
RWGFILTER	Software for synthesis and design of ridged waveguide filters by computer optimization
RWGMODEL	Software for electromagnetic modeling of ridged waveguide discontinuities
RWGSYNFIL	Software for synthesis of ridged waveguide filters
SUN	SUN workstation
TE	Transverse electric
TEM	Transverse electromagnetic
TM	Transverse magnetic
UMTS	Universal mobile telecommunication system

WG-16	22.86–10.16 mm
WLAN	Wireless local area network
WR90	22.86–10.16 mm
X-band	8.2–12.4 GHz

About the Author

Djuradj Budimir was born in Serbian Krajina, formerly Yugoslavia. He received Dipl. Ing. and M.Sc. degrees, both in electronic engineering, from the University of Belgrade, Belgrade, Serbia, and a Ph.D. degree in electronic engineering from the University of Leeds, Leeds, UK. In March 1994, he joined the Department of Electronic and Electrical Engineering at King's College, University of London. Since 1997, he has been with the Department of Electronic Systems at the University of Westminster in London. His research interests include analysis and design of hybrid and monolithic microwave integrated circuits such as amplifiers and filters, dielectric resonator filters employing high-temperature superconductors for communications systems, the application of numerical methods to the electromagnetic field analysis of passive microwave and millimeter-wave circuits, and the design of waveguide filters and multi-plexing networks for microwave and millimeter-wave applications. Dr. Budimir has published over 30 technical papers in the field of microwave CAD and owns a consulting company.

Index

The Artech House Microwave Library

TRAVIS Pro: Transmission Line Visualization Software and User's Manual, Professional Version, Robert G. Kaires and Barton T. Hickman

TRAVIS Student: Transmission Line Visualization Software and User's Manual, Student Version, Robert G. Kaires and Barton T. Hickman

Yield and Reliability in Microwave Circuit and System Design, Michael Meehan and John Purviance

For further information on these and other Artech House titles, including previously considered out-of-print books now available through our In-Print-Forever™ (IPF™) program, contact:

Artech House	Artech House
685 Canton Street	Portland House, Stag Place
Norwood, MA 02062	London SW1E 5XA England
781-769-9750	+44 (0) 171-973-8077
Fax: 781-769-6334	Fax: +44 (0) 171-630-0166
Telex: 951-659	Telex: 951-659
e-mail: artech@artech-house.com	e-mail: artech-uk@artech-house.com

Find us on the World Wide Web at: www.artech-house.com